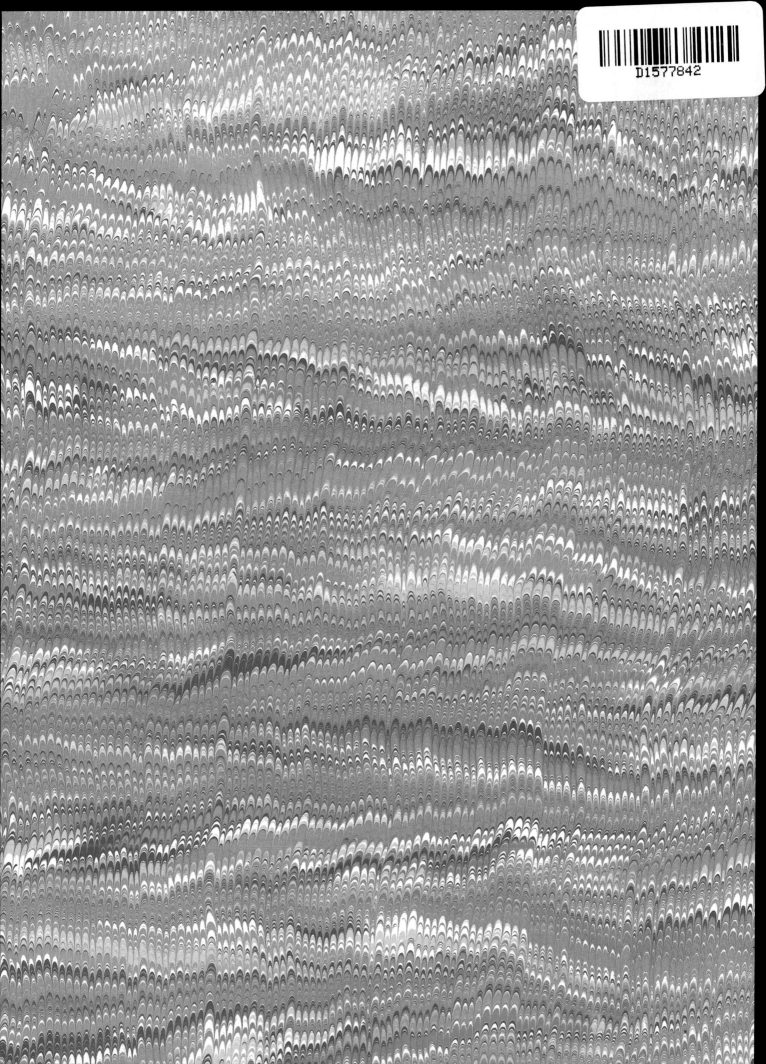

More
Collectible Bells:
Classic to Contemporary

Donna S. Baker

4880 Lower Valley Road, Atglen, PA 19310 USA

Dedication

For my nephews, Bryan and Evan.
You make "aunt-hood" a joy.

**Library of Congress Cataloging-
in-Publication Data**

Baker, Donna S.
    More collectible bells: classic to contemporary/
    Donna S. Baker.
      p. cm.
    Includes bibliographical references and index.
    ISBN 0-7643-0865-3 (hardcover)
    1. Bells--Collectors and collecting--United States--
Catalogs. I. Title.
NK3653.B36 1999
681'.868848'075--dc21   99-17656
                                 CIP

Designed by "Sue"
Type set in University Roman Bd BT/GoudyOISt BT

ISBN: 0-7643-0865-3
Printed in China
1 2 3 4

Published by Schiffer Publishing Ltd.
4880 Lower Valley Road
Atglen, PA 19310
Phone: (610) 593-1777; Fax: (610) 593-2002
E-mail: Schifferbk@aol.com
Please visit our web site catalog at **www.schifferbooks.com**

This book may be purchased from the publisher.
Include $3.95 for shipping.
Please try your bookstore first.
We are interested in hearing from authors
with book ideas on related subjects.
You may write for a free catalog.

In Europe, Schiffer books are distributed by
Bushwood Books
6 Marksbury Rd.
Kew Gardens
Surrey TW9 4JF England
Phone: 44 (0)181 392-8585; Fax: 44 (0)181 392-9876
E-mail: Bushwd@aol.com

Contents

Acknowledgments 4

Introduction 5

Chapter One: Bells of All Kinds 7

Chapter Two: A Medley of Materials 22

    Metal 22

    Glass 32

    Ceramic 48

    Enamel 55

    Wood 58

    Mixed Materials 61

Chapter Three: Purposeful Bells 65

    Religious and Inspirational 65

    Animal Bells 68

    Call Bells and Doorbells 75

    Rattles and Toys 81

Chapter Four: Beauteous Bells 85

    Souvenir Bells 85

    Commemorative and Convention Bells 89

    Figurals and Figurines 95

    Historical People 112

    Farmyard to Forest—Figural Animals 115

    Sarna Bells 123

    Miniatures 126

    Musical Bells 127

    Seasonal and Holiday 129

    Jewelry 143

Chapter Five: Melodious Motifs 148

    Birds 148

    Flowers 155

    Children 167

    Novelty Clappers 175

Chapter Six: Bells and Beyond: Multi-Purpose and Kindred Spirits ... 177

Appendix: The American Bell Association 188

Bibliography and Recommended Reading 189

Index 191

Acknowledgments

A book like this does not happen without the cooperation and help of many dedicated people. For sharing their wonderful bell collections and bell knowledge with me, I am deeply grateful to Sylvia A. Nemecek, Jane Connor, Margaret Smith "Penny" Wright, Jean Cline, Amy Haggblom, Blair Loughrey, and Tim and Kim Allen. The contributions and generosity of all are deeply appreciated.

Rita Walker, long-time member of the American Bell Association and one of the organization's past presidents, provided the values included in the book. I am most grateful to her for helping with this important facet of the project and for graciously sharing her experience and expertise.

Once again, the support and encouragement from members of the American Bell Association has been most gratifying. I feel privileged to have met and talked with so many exceptional members of ABA, and their collective enthusiasm is reflected within the pages of this book. Bell collectors have a wonderful resource in this organization!

My colleague Jennifer Lindbeck ably assisted with the data recording and photography. My thanks to her as well as to the editorial and design staff at Schiffer Publishing for helping to make this book a reality.

Introduction

Look around your home—on the bookshelves, windowsills, mantles, or end tables. Check the garden, jewelry drawer, perhaps the children's toy chest. Do you see a bell? If you are already a bell collector, chances are you see more than you can count! But even if you are not, no doubt a bell or two resides somewhere in your home, perhaps a souvenir from some long ago excursion, a thoughtful gift, or simply a welcome addition to the household decor.

For some, that resident bell is a pleasant, if not particularly engaging, aspect of their environment. Yet for others—most likely those holding this book in their hands—that bell and all others hold a deep fascination, a charm and attraction that extends from the daintiest of tea bells to the sturdiest of cowbells, a desire to be studied, reflected upon, enjoyed, and appreciated. If you fall into that latter category, you will hopefully be captivated as well as enriched by the medley of bells shown on the pages that follow!

Bells have played a prominent role in our society since the earliest of civilizations. As one collector notes, "[t]hey have been the subject of veneration, superstition, symbolism, emotion and influence." (Hamlin 1995, 24) Their sounds stir a host of emotions, from joy and merriment to sadness and alarm; all are dramatically—perhaps even melodramatically—conveyed in Edgar Allen Poe's famed *The Bells,* a poetic and often eloquent rendition of bells' evocative powers.

For collectors, both antique and contemporary bells are desirable. While it is especially satisfying to study the historical significance of old bells—particularly those that originally served some pragmatic function—modern bells often compensate for their youth with a beauty, artistry, or creativity of great proportion.

Like its predecessor, *Collectible Bells: Treasures of Sight and Sound,* this book is designed to illustrate a broad, but by no means all-inclusive, array of bells. For ease of comparison, it has been organized in the same basic manner as the former—yet the contents are entirely new. Indeed, the wonderful diversity of bells makes it possible to illustrate the same broad categories with completely different bells. As before, this book starts with chapters on the primary forms of bells and the materials used to make them, then divides the majority of remaining bells into those intended for a specific purpose and those intended primarily as decoration. Additional chapters highlight special motifs enjoyed by collectors as well as multi-purpose bells and "kindred spirits"—objects shaped like bells that are used in other ways.

As you turn the following pages, remember that by today's common wisdom you are a collector if you have but three or more of the same item. Together with this book, those bells you found as you gazed around your home may lead you on a new and most enjoyable journey!

About the Values

The values listed for the bells in this book are intended to provide readers with a general idea of what they might expect to pay for the same or similar bell in today's market. The values represent a guideline only and are not meant to "set" prices in any way. It is entirely possible to purchase a bell for a higher or lower amount than the value shown here, as many factors affect the actual price paid. These factors include the bell's age, condition, workmanship, size, and scarcity. None of these are absolutes, however. Older or larger bells, for example, are frequently among the most expensive, yet some contemporary bells of relatively small size may command high prices due to their exceptional quality and limited availability. In addition, geographic location, the context in which the bell is sold, and the buyer's relative desire to own a particular bell can significantly impact final purchase prices.

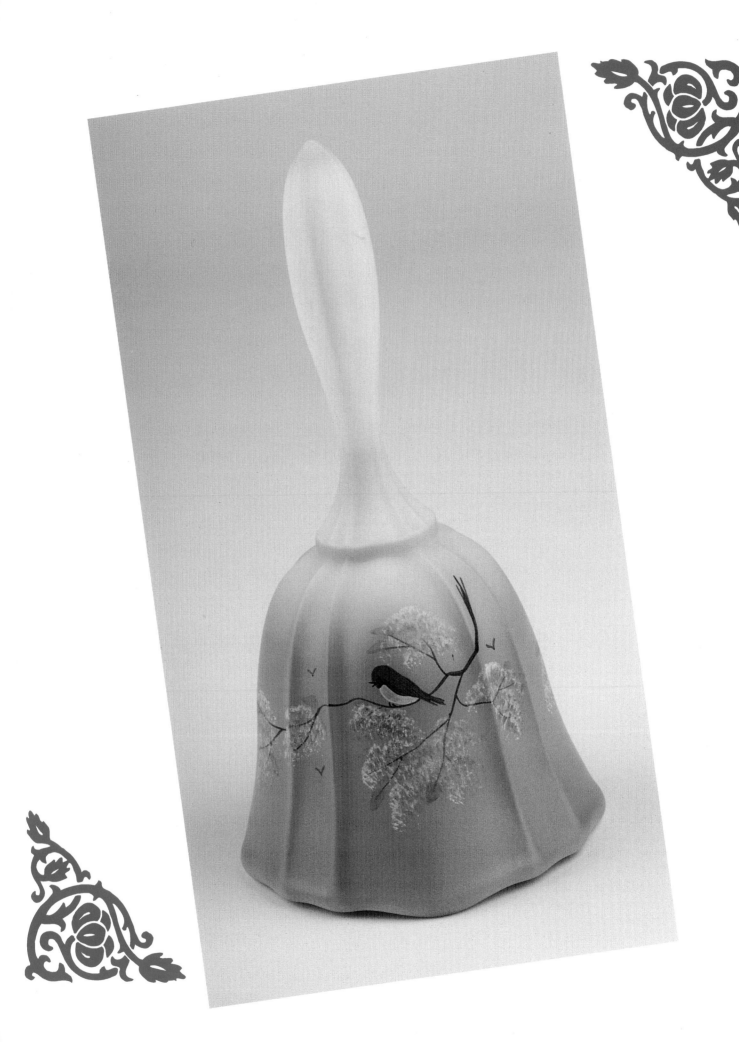

Chapter One

Bells of All Kinds

Those new to the world of bells do not always realize the scope of objects encompassed by this broad heading. As one begins learning and exploring, however, it becomes quickly apparent that bells can assume many forms, fashions, even names. Indeed, while the term "bell" can be applied at least generically to all the items pictured in this book, several other terms are more appropriately used at times.

Similar to many forms of art, music, and culture, the reason for this lies partly in the dual development of bells in both the East and the West. In the West, bells primarily evolved into the characteristic "open mouth" or inverted cup shape: a straight or slightly concave base flared out at the rim and topped by an elongated handle, producing sound through the action of a clapper suspended inside. While the size of the bell, the material used, and the degree of ornamentation vary widely, this traditional style is still the one most people visualize when they hear or read the word "bell."

Traditionally shaped copper bell from Turkey with elaborate flower design on the base and graceful metal handle. 5.25" high. $25-30.

Porcelain, open mouth bell with gold tipped handle and slender gold trim. Unmarked, but has traditional "Gibson Girl" painted on the base. 5.5" high. $15-20.

Very feminine pink and white frosted glass bell in open mouth shape, trimmed with lace, white bow, and cameo. Label inside reads "PG" with letters intertwined. 4.75" high. $15-20.

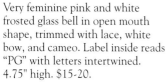

Below:
Bohemian crystal bell in gold-flecked white, traditional shape with gold overlay heart design. 5.5" high. $15-20.

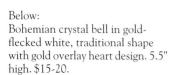

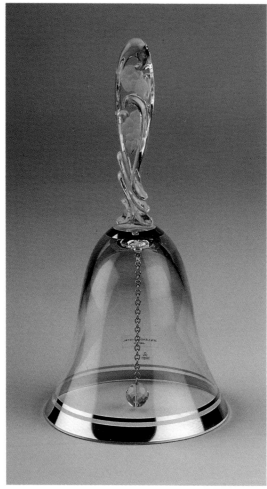

Contemporary crystal bell from the Danbury Mint with open mouth shape and floral motif handle. 7.75" high. $30-40. *Courtesy of M. "Penny" Wright.*

Purchased in West Virginia, this open mouth, very lightweight bell is crafted from coal, as is the clapper inside. 6" high. $20-25.

Gongs, on the other hand, owe their heritage to the Eastern world. Although they eventually found their way to Europe and America as well, gongs most likely originated in China and are characterized by a flat, round shape, producing sound when struck in the center by a mallet or hammer.

Gongs are typically suspended from a separate, often decorative frame and their tones are deep and sonorous. Some gongs do share a cup shape with their Western cousins, but do not have a clapper inside—like the disk shaped gongs, they require a mallet or hammer to "ring."

Brass gong with small bell suspended from the "beak" of a stylized crane, separate mallet for striking. Marked "China" on the bottom of the small rectangular base. 7.5" high. $10-15.

Brass gong with two small bells suspended on either side of a vase shaped frame, separate mallet for striking. "China" is stamped on the back of the frame. 6.5" high. $15-20.

Brass gong with frame comprised of two cranes, small bell suspended from a hook that attaches to the scroll like decoration at the top. No mallet attached. Marked "China" on the bottom of the base. 7" high. $20-30.

Brass gong with four small bells suspended on frame, two birds with crossed wings form top of frame. Separate mallet. Overall height: 8". Each bell: 1.25" dia. $25-35. *Courtesy of M. "Penny" Wright.*

A similar set of gongs, with slightly different birds. Overall height: 7" high. Each bell: 1.25" dia. $25-35. *Courtesy of M. "Penny" Wright.*

Brass gong suspended from a frame comprised of two stylized fish, separate mallet attached by small round holder. Overall height: 5.5". Bell: 1.5" high. $20-25.

Chinese gong on pagoda type frame, with jade inset above bell. "Made in China" on bottom cross bar of frame. 7.5" high. $40-50.

Brass gong from China hanging from rectangular frame. On one end of the unusual mallet is an amber colored stone with a filigree setting. The inside of the bell is marked "China." 10" high. $35-50.

Brass gong from China hanging from oriental torii frame. Marked "China" on both the bell and the frame, mallet not original. 4.25" high. $30-40.

Brass gong held by two pointy-eared pixies sitting on toadstools, each holding a lantern in the other hand. The wooden base has two small holders in the front to support the large mallet made of bamboo. Although probably made in England, this gong was purchased by its current owner in Canada. 7" high. Base: 11" wide. $40-50.

Opposite page:
Oriental dinner gong, consisting of a kimono-clad male figure supporting an inverted bowl gong with one hand and holding a long staff in the other, c. 1900. The gong may have originally been held in the more usual downward position via a support between the bowl and figure. Original leather covered mallet. 10.5" high. Gong: 5.5" dia. $75-85.

Detail of figure holding the gong shown on previous page.

Heavy old bronze temple gong from Japan, called a *wani guchi* or "shark's mouth." It has a raised flower decoration in the center but no mallet. 6.5" high. $90-100.

Elegant, rare brass gong engraved with bird and leaf motif. The gong is supported by the bent figure of a man, while the heads of three other men appear on the gong's ornate frame: one at the apex and two on each side. The base appears to be made of silver-plate over brass and has no markings. The dealer from whom this bell was purchased placed its origin as England. The wooden mallet is not original to bell. 11.5" high. $75-85.

Detail of the gong and frame at left.

Unique brass gong suspended from two large horns mounted on an oak base. The horns are reported to be from Blue Kerry cows and the entire piece was supposedly owned by the British Earl of Mt. Batten. 9" high. Base: 14" wide. $50-75.

This unadorned silver gong rests in a circular frame that mirrors the gong's shape and stands on four round ball "feet." Purchased near Donegal, Ireland, the gong has Irish silver marks on the bottom of the frame. 9.25" high. $35-50.

Indian ankle bracelet adorned with an abundance of tiny crotal bells. The bells are attached in groups to small metal loops surrounding the perimeter of the bracelet. 3" dia. $25-30. *Courtesy of M. "Penny" Wright.*

The term "crotal" may be unfamiliar to some, but it too applies to bells: crotal bells are those which are completely enclosed, usually spherical in shape, and produce sound via small pellets contained within. Jewelry, ornaments, and children's rattles are typical examples of crotals, as are the well-loved metal sleigh bells that "jingle" so appealingly.

Dancer's toe ring from India, ornamented with small crotal bells, a tiny mirror, and decorative stones. 1.5" dia. $15-20.

These two colorfully painted doll bells are from Russia. Crotals both, they are completely enclosed on the bottom. 3.75" and 5.25" high. $20-25 for pair.

Happy faces characterize these two colorful crotals, one a pink bunny, the other a red-haired clown. The clown crotal is one of several made by a past president of the American Bell Association (ABA) who gives clown performances and calls herself "Bella the Clown." The bells were given as small gifts to other members of the group. Bunny: 1" high. Clown: 1.25" high. $5 each.

These two brightly decorated crotals hail from Japan. On the left is a pear shaped crotal featuring red and white flowers and a beige tassel; it comes from Kyoto. 2.75" high. On the right is a round ceramic bell with fancy rickshaw and flowers on the front, red braided cord tied to the handle. A label on the bottom reads: "Yamayo / Japan." 3" high. $10-15 each.

Silver-plated twist bell set on round marble base, acorn shape decoration around lower portion of bell. 5" high. $85-95.

Among the most intriguing of bells are those that fall under the category of mechanical. Also called tap bells or twist bells, such bells use a mechanical striking device of some kind to produce their sound—ringing the bell may require such actions as tapping or twisting a knob that sits on top, twirling an axle that holds the bell, snapping a lever attached to the side, or depressing some part of the bell to activate a wound-up spring. While these actions—as well as the very term "mechanical"—may sound rather humdrum, in reality these bells are often anything but. Frequently used in Victorian era hotels, restaurants, or homes, they combined usefulness with comeliness and may be appreciated for their artistry as much as for their competency.

Unusual silver tap bell with figure of barefoot girl standing on oval platform, one hand on a long rod which forms part of the handle for the bell. Marked "Meridan B. Company," patent dates Aug. 1863 and April 1856. 9.5" high, 6.5" wide. $125-150.

Brass tap bell with side clapper mechanism, a reproduction of an earlier bell. 5.25" high. $60-75.

Detail of the clapper mechanism on the brass tap bell.

Very old French slave call bell decorated with three abalone shells, clusters of brass grapes, and brass crescent finial. This mechanical bell is rung by depressing the metal loop at the top; this lifts the metal lever with ball on end and strikes the bell. 4.25" high. $175-195.

Metal chime on wooden base with separate mallet. When struck, this chime rings for a long time with an especially lovely tone. 5.75" long. $15-20. *Courtesy of M. "Penny" Wright.*

Chimes and windbells are still other forms of bells. Chimes, more often associated with the music world, are sets of bells that are tuned musically and produce a very melodious, pleasing tone. Windbells are meant to hang—they usually contain a long, separate clapper that knocks against the bell as both swing pleasantly to and fro in the wind.

This set of tubular dinner chimes, said to be from the eighteenth century, originally hung from the separate wooden stand and was struck with the accompanying mallet. The mallet is 9" long with a wooden handle and a metal head; the chimes are graduated in length, from 4' 1" long to 4' 8" long. $45-50. *Courtesy of M. "Penny" Wright.*

Long, cast bronze windbell by Paolo Soleri, an Italian architect who came to the United States in 1947 and founded Arcosanti, an "urban laboratory" in Scottsdale, Arizona. Proceeds from the sale of bells cast at Arcosanti help fund the ongoing project, which is designed to be an energy-efficient community combining architectural and ecological concepts. The bell measures 25" long from end to end. $90-95. *Courtesy of M. "Penny" Wright.*

Pottery windbell in shades of blue, brown, and gray, marked "Mexico" inside. 11.5" long. $10-15.

Unglazed pottery windbell with metal clapper, done in earthtone colors with cactus decoration on front. Made by Western Potteries, Scottsdale, Arizona. 13.5" long from end to end. $15-20. *Courtesy of M. "Penny" Wright.*

Ceramic windbell with hand painted farm scene, copper clapper. 11" long. $10-15.

Chapter Two

A Medley of Materials

One logical way to categorize bells is by the material used to make them, an aspect which certainly affects each bell's overall appearance. Accordingly, this chapter highlights the most common materials employed, including a section on bells with dual personalities, i.e., those made from a combination of materials. Each material has its own unique look and feel, qualities which may attract one collector more than another: some love the sturdiness and weight of old metal, others prefer the airiness and grace of glass, still others favor glossy, rich enamel.

Metal

Of all materials used for bells, metal is by far the most durable. Metal bells, consequently, have withstood the dual tests of time and weather far better than any others. When unearthed, ancient bells are invariably made of metal, and while their appearance may have deteriorated through the years, their very existence is testament to the enduring properties of this substance. Similarly, hard-working bells that are regularly exposed to the elements—farm bells, school bells, or ship's bells, for instance—are almost always made of metal.

Metal's durability, however, in no way detracts from its versatility. One need only visualize the contrast between a massive bronze church bell and a dainty silver tea bell to appreciate the divergent ways in which metal's qualities enhance a bell. And when it comes to sound—that all important aspect of bells—very little compares with the beautiful timbre and resonance that emanates from a well-made metal bell.

Bells fashioned from metal are most typically produced through casting, a process in which the molten metal is poured into a mold and allowed to solidify. Brass and bronze, both alloys of copper, are two of the most frequently used metals. Brass is made from copper and zinc, bronze from copper and tin; when old, these two metals can be difficult to tell apart. The characteristics of bronze make it a favorite medium for sculptors of bells and other objects:

it melts more quickly than other metals, expands as it cools (thereby flowing into every crevice of the mold), and can be easily embellished with tools once hardened. As bronze ages, oxidation causes its surface color to take on the green or gray-green hue called a patina. This is usually considered desirable, adding to the object's authenticity and overall fine appearance.

Two brass bells from China with traditional designs. On the left is a tortoise shaped bell; it is marked "China" and bears Chinese characters on both sides of the half oval handle. In China, notes Lois Springer in *The Collector's Book of Bells*, the image of a tortoise represents strength, endurance, and long life; Springer further notes that bells like this one were often used as christening gifts for children. 3.5" high. $20-25. The bell to its right has a handle shaped like a pagoda and is also marked "China." 4.5" high. $15-20.

Small brass bell in the shape of a Dutch windmill. 2.5" high. $15-20.

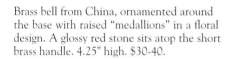

Brass bell from China, ornamented around the base with raised "medallions" in a floral design. A glossy red stone sits atop the short brass handle. 4.25" high. $30-40.

Opposite page:
Old bronze Oriental bell with long cylindrical clapper, possibly from Thailand. A second bell may have once been attached to the clapper, which has a hole at its very end. The clapper is attached to the bell by a chain that extends through an opening just below the handle. 5" high. $20-25.

Heavy brass Chinese bell with figural rooster handle. The blue, red, and green enameled decorations around the middle and top of base are similar to those on the bell at left. 4" high. $35-50.

Old bronze Italian Renaissance bell with figural handle of seated cherub. Finely detailed flowers and faces in bas relief encircle the base. The acorn shaped clapper features miniature decoration halfway up its interior. 5.5" high. $150-175.

Old brass dinner bell with graceful shape and handle, originally purchased in London. 3.5" high. $20-25.

The acorn shaped clapper from the Italian Renaissance bell.

Art Nouveau style triple brass bells with greenish hue, marked "Made in Italy." The three bells are meant to resemble flowers, complete with multiple brass leaves. 5" high. $20-25.

Heavy antique bronze bell with bulbous handle, accented with concentric lines covering the base and handle. The Irish dealer from whom this bell was purchased in 1973 reported that it had been used as a church bell, although its actual function there is not certain. 6.25" high. $60-75.

Above and right:
Indian scrubber bell, also known as a masseurs' bell. Punchwork scrubber bells, typically decorated with animals on their handles, were originally used in India for scrubbing mud off the feet of women working in the fields. A raised pattern underneath helped with the cleaning. This very old bell with dark patina has triangular shaped openings around the base and a typical scored bottom. The figure which serves as the handle appears to be a cat, though it may also be a rat. 3.25" high. $75-100.

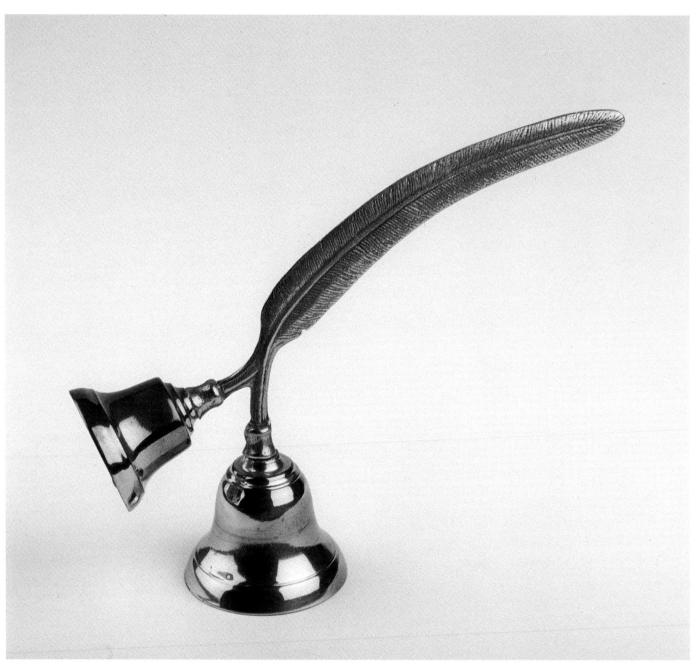

Double bell with shared feather quill handle, from England. 9"
long from tip of smaller bell to tip of feather. $15-20.

Brass bell with deeply fluted base and matching handle. The small black ball separating the base and handle provides an interesting contrast in shape. Purchased new in Bermuda in 1962. 4.5" high. $25-30.

Heavy metal bell with gong type shape, said to have been used as a trolley or street car signal bell. Such bells were mounted above the motorman's head with the open side facing up. They were operated by a cord from the rear platform so the conductor could signal stops and starts. This bell is missing its original mounting frame and has had a separate clapper added on. 10.5" dia. $10-15.

Small metal bell with intricate flower and leaf design, beige rope attached to handle. 2.5" high. $5-10.

Crudely cast metal bell with bas relief designs on base and cloverleaf shaped handle. 4" high. $10-15.
Courtesy of M. "Penny" Wright.

Large, heavy metal bell, riveted on one side. The bell is attached to a sturdy chain via a stirrup-shaped piece of metal; another "stirrup" is fastened to the other end of the chain. Although most likely an animal bell, this bell was also said to have been used as a slave bell, the stirrups (added on later) allegedly fitting around the slave's ankles. Bell: 8.5" high. $15-20.

One of the first metals used by humans, silver has retained its status as the metal of choice when elegance and refinement are desired. Along with gold and platinum, silver is considered one of the precious metals; its historical use as a standard of currency only increased its strong association with value—monetary and otherwise (in the U.S., expanding use of silver for industrial purposes led to the withdrawal of all silver coins from circulation by the late 1960s). Highly malleable, silver is used for silverware, jewelry, decorative objects, and, of course, for bells. Bells can be made of sterling silver (designated as 92.5 percent silver) or of silver plate; hallmarks, dates, or other notations can help ascertain the identity of an individual piece. These are not always easy to find, however, and may require a bit of sleuthing:

> Small objects like bells usually have fewer places where a mark can be discreetly stamped. Marks are frequently small, making it easy to overlook. Collectors should use a magnifying glass and thoroughly examine each piece; underside, rim, edges, and any recesses. (Pecor 1996, 35)

Sterling silver bell from Mexico with unusual handle and flowers embossed around the rim. Marked "Sterling / Mexico" inside base. 3.25" high. $30-35.

28

Silver-tone bell with decorative engraving around base and feather quill as handle. 5" high. $20-25. *Courtesy of M. "Penny" Wright.*

A trio of dainty silver table bells. From left: unmarked silver handbell with beaded edge around base and nice tone, appears sterling but may be silver plated. 3.75" high. $30-35; Art Nouveau silver-plated tea bell by Gorham, marked "Gorham 014" inside. 4" high. $35-40; sterling silver bell with coiled rope handle and scalloped edge, marked "Sterling 55" on front along with the monogram "C." 3.5" high. $30-35.

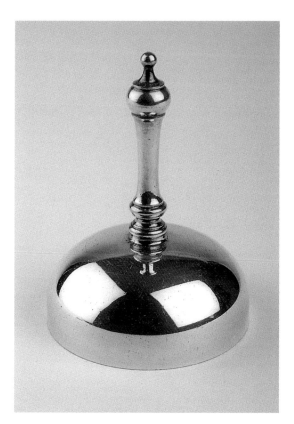

Heavy, silver-plated dinner bell with squat shaped base, c. 1870. 3.5" high. $35-40.

Twin silver bells, joined by one stem with silver "petals" adorning the top of both bells. Signed "Eberle" but no other markings. 6.5" high. $25-30.

Silver-plated bell with ornately decorated faces embellished by flowers on both sides of the tall handle. The sterling handle matches a flatware pattern. 5.5" high. $20-25.

Sterling silver bell with unusual twisted handle. Marked "Birks / Sterling / 1/ 41" on outside of base. 3" high. $30-35.

Although they have nothing to do with the composition of the metal, marks on the *outside* of some silver bells are interesting as well. Included here are several bells bearing monograms or inscriptions, some probably one-of-a-kind. Perhaps gifts or treasured keepsakes, each undoubtedly conceals a tale or two behind its shimmery facade. How fascinating it would be to learn the origin of these bells, and the identity of the long gone owners whose names or initials now tantalize us with their incompleteness!

Sterling silver handled bell by Whiting Sterling. The handle is engraved with the initial "M" on the front and has an enameled geometric design on the very top. 3.5" high. $35-45.

Silver-plated bell with large monogram on front and interesting twist handle, unmarked. 3.5" high. $35-40.

Sterling silver tea bell with swirled repoussé design on base, c. 1891. Probably a gift, the bell has the word "Mother" inscribed on one side of the handle and "Dec. 25, 91" on the other. 3.75" high. $40-50.

Elegant old silver bell from England with floral embossing around base and monogramming on front. Six-sided handle engraved with interesting designs. Inside marked: "REG / 380 / S." 4.75" high. $50-60.

Perhaps a favor from a long ago wedding or anniversary celebration, this silver bell from c. 1910 is engraved "Bonnie & Stephen" on the flat portion of its tall handle. The reverse of the handle is marked "Webster Sterling." 4" high. $15-20.

In contrast to their metal counterparts, the charm of glass bells is largely derived from their very fragility—their color-stained, often ethereal essence, their delicate, smooth contours, their tinkling ring reminiscent of the most palatial dining rooms.

To create glass objects, hot, liquid glass is given form in a number of ways, including hand blowing, mold blowing, pressed glass, and machine-made techniques. The glass is then slowly cooled to protect its integrity, after which patterns can be added through such processes as cutting, engraving, or etching. Special effects enhance and add interest to the final product. These include techniques such as overlay glass (using two or more layers), Millefiori (applying tiny pieces of colored glass to the basic glass shape while it is still hot), or carnival glass (spraying the hot glass with metallic salts to create a showy, iridescent finish).

Featured here are bells representing a mini-tour of glass shapes and styles from a host of different countries, including Germany, Hungary, Poland, Bohemia, Italy, Ireland, Great Britain, and the United States. They incorporate bells both clear and opaque; bells with intricate, cut glass designs; bells in traditional colors and designs; bells with hand painted, often artistic designs; bells in clear, classic crystal. A number of different companies are also represented, many with names world renowned in the field of glass and other artworks.

Hand cut lead crystal bell of amethyst and clear glass from West Germany, made by E & R "Golden Crown." Narrow gold trim around rim, floral decoration in panel above. 5.75" high. $35-45.

Two hand cut lead crystal bells, one in purple and one in green, both with crystal handles. The bell on the left is from the "Hofbauer Collection" and is 8" high. Its companion is marked "Genuine Lead Crystal Made in West Germany" and is 6" high. Left: $30-35. Right: $20-25.

Cut lead crystal bell in amber with grape and flower etched design, crystal handle. Made by the House of Goebel, West Germany. 7.75" high. $35-40.

Tall glass bell made in Hungary, clear crystal with rose colored middle, gold bands, etched leaf design, and swirl handle. 9.25" high. $25-30.

Cobalt blue, etched crystal bell from Poland, with clear crystal handle and crystal bead clapper. Purchased new in 1977. 5.75" high. $20-25.

Detail of the trademark and artist's signature on the Rosenthal bell.

Glass bell by Rosenthal, designed by the Danish artist Bjorn Wiindblad. The bell has an angel design on the front, with the Rosenthal trademark and artist's signature on the reverse. 4.25" high. $20-25.

Two bells made of Bohemian crystal, one in rose and the other in cobalt blue. Both have gold thistle designs around the base and are trimmed with bands of gold around the rim and just below the handle. Bohemia, a province in Europe that is now part of the Czech Republic, is well known for its fine quality glass, much of it exported to the United States. Both 7" high. $30-35 each.

A trio of Bohemian crystal bells, all in cobalt blue with varying gold overlay trim and designs. Average height: 7". $30-35 each.

Two Millefiori glass bells, both made in Murano, Italy. The bell on the left has a small loop handle and is 6.25" high. $65-70; the smaller bell on the right has a dolphin shaped handle and is 5.25" high. $45-50.

A third Millefiori glass bell, with over one hundred brilliant beads fused in clear glass to form a continuous flower design. The clear twisted handle is flecked with gold and the clapper is a faceted bead. Red label on back reads: "Murano Glass Italy." 5" high. $65-70.

This ribbon glass bell from Murano, Italy, has a bulbous handle and features multi-colored twisted "ribbons" alternating with panels of tiny white lattice on the base. 4.5" high. $50-60.

Blue and gold glass bell from Murano, Italy with similar swirled ribbon effect and gold loop handle. Red label on back reads: "Made in Murano Italy." 4.4" high. $45-50.

36

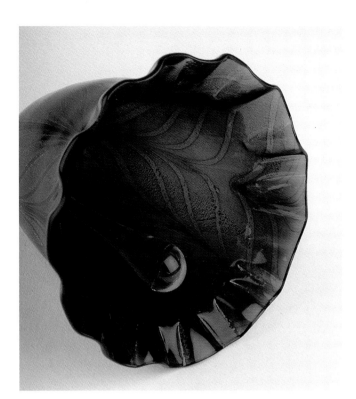

The deep green color of the glass is best seen from the interior of the bell at left.

This glass bell from Italy is deep green, overlaid with a stylized gold leaf design. It has a clear glass looped handle and is 6" high. $50-60.

Old Venetian glass bell in pale green with clear swirl handle. The clapper is a small bead of green glass. 5" high. $20-25.

Ruby glass bell from Murano, Italy, with a gold handle and gold trim. A continuous hunting scene encircles the central portion of the base; above and below are white enameled decorations that contrast strikingly with the deep crimson glass. 5.75" high. $45-50.

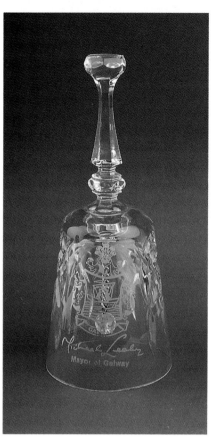

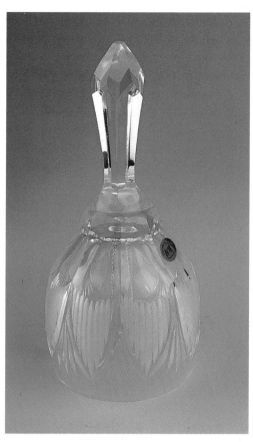

Glass bell from Steinbock, Austria, deep blue with snowflake design and gold metal handle. 4.25" high. $40-45.

Lead crystal bell by Galway. The city seal of Galway appears on the front, along with the signature of Mayor "Michael Leahy." The back has a small green tag reading "Galway Lead Crystal." 5.75" high. $40-50.

Hand cut crystal bell in "Clonemacnoise" pattern, made by the Tyrone Crystal Co. of Ireland. Panels of cut frosted glass encircle the base. 7.5" high. $75-80.

Although world renowned for their ceramic production, the Wedgwood company has also produced glassware. These two crystal bells by Wedgwood both feature cameo designs on the front made of the more traditional jasper material. The bell on the left is amethyst crystal with a "Queen Elizabeth" cameo in beige and white. It was made to commemorate the 25th Anniversary of Queen Elizabeth II's coronation. The bell on the right was Wedgwood's first crystal bell; it was made in a limited edition of 500 to commemorate the XXI Olympiad in Montreal, July, 1976. The bell's blue and white jasper cameo illustrates the Olympic torch runner. Both 6.25" high. $60-70 each.

Right:
Clear crystal bell with twist handle and sketch of Belvoir Castle, a stately mansion near Leicestershire, England that serves as home for the Duke and Duchess of Rutland. The castle has been used as the location for several film and television programs. 7" high. $35-40.

Far right:
Cranberry glass bell from England by Royal Doulton, with clear crystal handle and Royal Doulton hallmark on back. 5" high. $35-40.

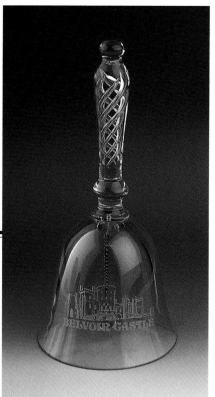

Among American companies, Fenton Art Glass has distinguished itself as one of the most prolific manufacturers of glass bells. Founded in 1905 by Frank Leslie Fenton, the company flourished in Williamstown, West Virginia, and produced its first bells in 1910-1914: these were souvenir bells in carnival glass made for the BPOE Elks conventions in four different cities. The company made several more bells during its early years, but production was discontinued in the thirty years between 1937 and 1967. The 1970s saw a resurgence in Fenton bells, coinciding with a great increase in their popularity among collectors. Most Fenton shapes are given specific names, such as Faberge or Paisley, and each shape can be produced in a variety of colors or different types of glass. Still other Fenton bells are named for their pattern or decoration; the "Spring Flowers" bell on page 44 (featuring a scene from one of Norman Rockwell's famous paintings) is an example. Additional Fenton bells can be found elsewhere in this book: see the sections on commemorative bells, musical bells, and special motifs.

Three glass bells by Fenton in the Whitton shape, varied colors and rims. From left: red carnival glass with scalloped rim; Cameo Opalescent with flat rim; Velva Rose iridescent stretched glass with scalloped rim. 7" high. $30-35 each.

Glass bell by Fenton, Hobnail shape in Persian Blue Opalescent. First produced by Fenton in 1967, Hobnail bells have undergone many style variations. This bell is one of the later versions, introduced in 1989. Its double crimped crystal edge is known as Silver Crest. 5.75" high. $35-40.

Fenton originally used metal clappers on their glass bells, but replaced them in the 1970s with glass beads hung from silver chains. Bell collectors Ralph and Rita Walker helped initiate this change by visiting the Fenton company in the early 1970s and convincing one of the owners that the bells deserved more attractive clappers. The changeover was begun in 1975 and completed by 1977. This photo shows the two types of clappers, metal on the left and glass on the right. The bell on the right also illustrates the typical artist's signature found inside most Fenton bells.

Blown glass bell by Fenton, "Spiral" in cranberry glass. This bell has an inner layer of gold ruby, which is encased in clear crystal and then mouthblown into an optic mold for decoration. 6.5" high. $35-40.

Blown glass bell by Fenton, "Colonial Scroll on Royal Purple." This bell is from the limited edition Family Signature Series, comprised of bells representing some of the Fenton artisans' finest work. It was designed by Frances Burton, signed by Don Fenton, and hand painted by V. L. Anderson. The gold lettering which can be seen on the inside front is unique to this individual bell: it was personalized as a gift to the owner upon her installation as president of the American Bell Association in 1998. 6" high. $45-50.

Left:
Glass bell by Fenton, "Butterfly" in red carnival glass. The bell obviously gets its name from the colorful butterfly poised atop its handle. 6.75" high. $25-30.

Right:
Glass bell by Fenton, "Temple Bell," also in red carnival glass. The shape of this bell is formed by stylized lilies that cascade down the sides of the bell. 6.75" high. $20-25.

Below:
Glass bell by Fenton, "Beauty" in cobalt blue carnival glass. 6.5" high. $30-35.

Below right:
Detail of the Beauty bell, showing 1980s style Fenton logo just below the handle.

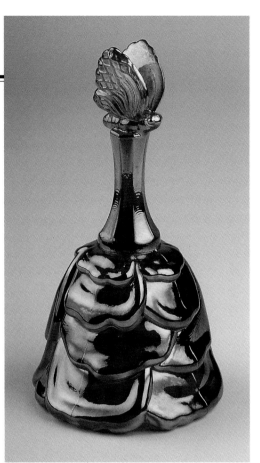

Glass bell by Fenton, "Sable Arche" in French Opalescent. This bell was introduced in 1980 and was one of Fenton's "Collecti Bells." The Collecti Bells were made from molds previously owned by a dinnerware company where they had been used to make goblets. 6" high. $25-30.

Glass bell by Fenton, "Threaded Diamond Optic" in Wisteria. This shape was introduced in 1978. 5.75" high. $25-30.

Left:
Glass bell by Fenton, Faberge shape in Pearl Pink. Fenton created the Faberge shape in 1976. 7" high. $30-35.

Pair of glass bells by Fenton, "Cross," in Shell Pink and Teal Royale. This bell, which has a scalloped lip and lilies decorating the base, was introduced in 1989 and also made in Crystal Velvet. 6" high. $20-25 each.

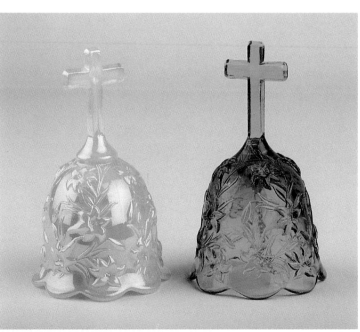

Glass bell by Fenton, "Basketweave," in lustrous iridized opal with primrose design on front. 6.25" high. $20-25.

Glass bell by Fenton, "Bleeding Hearts," designed by Robin Spindler. This bell is from Fenton's 1998 Designer Series and is #571 from a limited edition of 2500. It is in the Aurora shape and made of Burmese glass. 7" high. $45-50.

Glass bell by Fenton, "Copper Rose" in Paisley shape. The glossy black bell is decorated with a hand painted floral design and accented with copper. 6.5" high. $35-40.

Glass bell by Fenton, Paisley shape in Velvet Green. 7" high. $25-30.

Glass bell by Fenton, "Spring Flowers," featuring Norman Rockwell's popular porch scene on Custard Satin glass. Marked inside: "Spring Flowers / hand painted by Sue Jackson for River Shore Ltd. © 1982. No. 330." 7.25" high. $35-40.

Glass bell by Fenton, "Burmese Mariner's Bell," a limited edition from the 1986 Connoisseur Collection. A seahorse hovers over seashells in the hand painted decoration on the front. This bell is #375 from an edition of 2500. 6.5" high. $45-50.

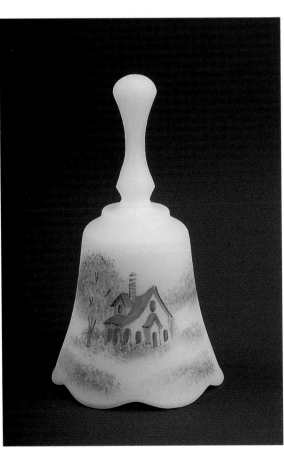

Petite glass bell by Fenton, "Cottage Scene" on Opal Satin, hand painted scene of thatched roof house and perennial garden. 4.5" high. $20-25.

Right:
Petite glass bell by Fenton, Heart shape in plum. 3.5" high. $15-20.

Far Right:
Petite glass bell by Fenton in Peach with diamond optic background and hand painted butterfly. 4.5" high. $10-15.

Below:
Trio of petite glass bells by Fenton. From left: "Tulips," Bow and Drape shape on opalescent glass with hand painted tulips. 4.5" high; "Pearly Sentiment," Heart shape on iridescent opal with porcelain bisque rose and tiny pink bow. 3.5" high; "Hearts and Flowers," pearl iridescent with scalloped lip and hand painted pink roses. 4.5" high. $15-20 each.

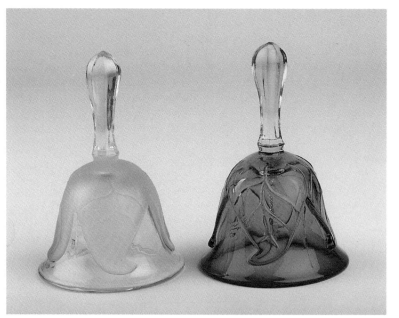

Pair of glass bells in pale yellow and deep orange with leaf designs, hand made by Viking. Both 6" high. Left: $20-25. Right: $25-30.

Petite glass bell by Fenton, "Bow and Drape" shape in Petal Pink. 4.5" high. $15-20.

Right:
Clear crystal, limited edition bell, by Fostoria, signed "F. 1978." A winter ice skating scene is etched on the frosted panel around the base. 7.25" high. $35-40.

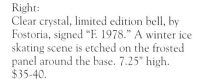

Carnival glass bell made by the Imperial Glass Company in "Peacock" hobnail pattern. Note how the hobnail pattern is extended onto both the middle and top portion of the bell's handle. Located in Bellaire, Ohio, Imperial—along with Fenton—was a major producer of carnival glass. They went out of business in the 1930s but later reorganized under a slightly different name and acquired the molds from several other former glass companies, such as Cambridge and Heisey. After several corporate changes in ownership, Imperial closed in the 1980s. 5.25" high. $25-30.

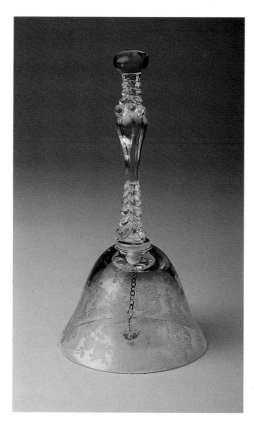

Etched glass bell in "Rose Point" pattern, made by the Cambridge Glass Company to match an extensive set of dinnerware and accessory pieces in the same pattern. 8" high. $35-40.

Clear glass Steuben bell from Corning, New York, with scroll or heart shaped handle, very artistic. First produced in 1938, this bell was designed by Samuel Ayres and the clapper is inscribed with the word "Steuben." Steuben was founded in 1903 and acquired by the Corning Glass Works in 1918. The company is known for its fine workmanship, creativity, and artistic rendering of glass. Only a small number of bells were produced, all highly collectible. 3.5" high. $150-175.

Hand blown crystal bell from 1994 by Tim Tiernan, of Lake Placid, New York. Graceful turquoise design accents the handle, shoulder, rim, and clapper. 7" high. $30-40.

Like an artist's canvas, ceramic objects start out simply but end up splendidly. Born of humble ingredients, they are transformed into works of beauty through a combination of heat, special glazes, and the talent of their creators. They offer bell makers the chance to explore a host of colors, shapes, textures, and styles, yielding final products with clearly defined and quite individual personalities.

The term "ceramic" is a rather generic one, referring to any object made of clay that is fired in a kiln. The specific qualities of the final object depend on several factors: the composition of the clay, the temperature at which it is fired, whether or not it is glazed, and the type of glazing used.

Earthenware, an opaque, porous form of ceramics, is produced at relatively low firing temperatures. Stoneware is similar to earthenware, but is water resistant and more durable; it is fired at a somewhat higher temperature. The highest temperatures are used to produce porcelain, also known as china. Porcelain was first developed by the Chinese around the ninth century AD and is both non-porous and translucent.

Several types of porcelain are used for bells and other decorative objects. Hard-paste or "true" porcelain contains only kaolin (a fine white clay) and feldspar (a naturally occurring mineral). Soft-paste or "artificial" porcelain was developed during Europe's early efforts to imitate the highly prized Chinese porcelain. It contains clay and glasslike ingredients and is fired at a lower temperature than true porcelain, remaining somewhat porous as a result. The most translucent form of porcelain is bone china, which originated in eighteenth century England and is made by adding bone ash to the ingredients for hard-paste porcelain.

All forms of ceramics can be either glazed or unglazed. Glazing is the form of decoration that gives ceramics its well-known smooth and glossy finish. It can be applied either before or after the piece is fired; if applied after the firing, a second firing is required in order to fix the glaze. Ceramic pieces that are left unglazed are referred to as bisque, or sometimes biscuit.

Graceful ceramic bell with aqua pearlized glaze and gold handle. 3" high. $10-15.

White porcelain bell trimmed with gold, bearing the initial "J" quite prominently on the front. Inscribed inside in gold: "With love from Aunt Mary." No other markings. 4.25" high. $20-30.

Cowbell shaped porcelain bell with gold trimmed handle and pastoral lake scene on the front. *Mer de Glace* (Sea of Glass) is located in the French Alps. 3" high. $45-50

Another cowbell shaped porcelain bell, this one from Germany. A windmill scene decorates the front, blue flowers, "Germany," and a crossed pipes mark are found on the reverse. 3.5" high. $45-50.

Most people are familiar with the works of Wedgwood, Britain's famed pottery dating from the eighteenth century. Josiah Wedgwood originated the concept of jasperware, an unglazed fine stoneware made in various shades and usually decorated with white bas relief figures. Of interest to bell collectors, however, is the fact that Wedgwood's pottery did not manufacture bells until the mid-twentieth century—and that their foray into the bell world began with five hundred blue and green bells specially commissioned for the 1964 convention of the

American Bell Association! Since then, the company has produced at least thirty different jasperware bells, including New Year's bells, Four Seasons bells, and Bicentennial bells, as well as several more in bone china and even glass.

Interior of the green jasperware bell, showing "Wedgwood / 1998" inscription.

Green jasperware bell by Wedgwood decorated with white cherubs in bas relief. Inside the bell in gold lettering appears "Wedgwood / 1998," inscribed by Piers Anthony Weymouth Wedgwood, the fourth Lord Wedgwood of Barlaston. Lord Wedgwood represents the company as an international ambassador and personally signed this bell as a gift to its current owner. 4.25" high. $65-75.

Lilac jasperware Four Seasons bell by Wedgwood. 4" high. $85-90.

Two additional bells by Wedgwood, both in blue jasperware. Left: "Floral Girl," made for the Danbury Mint's series called "Bells of the World's Great Porcelain Houses." 4" high. $50-60. Right: swan decoration, made for The Bell Collector's Club, an organization which also featured a series of bells from great porcelain houses. 3.5" high. $40-45.

Bone china bell from the Danbury Mint's series called "Bells of the World's Great Porcelain Houses, made in the United States by Gorham. 4" high. $30-35.

Hand painted pink and white porcelain bell with gold trim, marked "Limoges, France, Decor, Main" and believed to be old. Interesting, fan shaped brass handle nicely complements the flower and bow design on the base. 4.25" high. $45-55.

Two glazed porcelain bells by R.S. Prussia, pale green with delicate pink floral designs and wooden clappers. 3.5" high. $75-90 each.

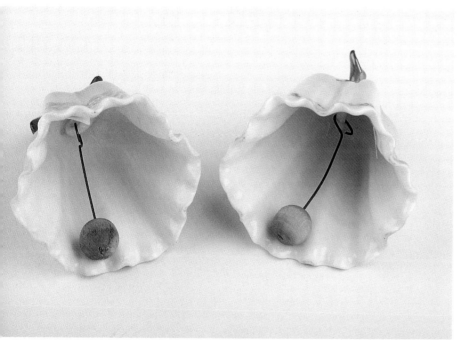

Interior of the R.S. Prussia bells, showing the wooden clappers.

Limoges china bell from France with gold ball clapper. The base is decorated with colorful insects, including a moth, bumblebee, cricket, and ladybug. Marked inside: "Ebeling & Reuss Co. / Limoges France." 3.75" high. $45-50.

Pale pink porcelain bell, trimmed with a cluster of gold grapes and leaves as handle. Made in Japan. 3.25" high. $10-15.

51

Glazed porcelain Sunbonnet Baby bell, "Friday," by the Royal Bayreuth Company of Germany, one from a series modeled after the designs of American illustrator Bertha L. Corbett. This bell is from a set of reproductions made c. 1977 and is marked on the inside: "Royal Bayreuth 1794 Germany." It is No. 517 from a Limited Edition of 1500. Friday represents sweeping. 3.5" high. $75-80.

Hand painted ceramic bell of Ukrainian origin, made by Daria Hanushevsky in 1993. The clapper is a painted wooden bead on a string. 5" high. $20-25.

Reverse of the Ukrainian bell, showing the artist's signature and date.

These two bells are both of Irish Dresden porcelain. After a German porcelain factory was destroyed in World War II, the company relocated to Ireland. The resulting products share characteristics from both countries. The "Lady's Bell" on the left is a cream colored bell with 14 kt gold trim, a green bow handle, and base decoration of a dancing ballerina with raised green lace skirt. 2.75" high. $40-45. On the right is a figurine with green cape and lacy ruffled skirt. Marked inside: "'Colleen Bell' Ireland." 3" high. $50-60.

"Spaghetti ware" bell with three small shamrock decorations. 3" high. $25-30.

Square shaped ceramic bell with flat round handle, bedecked with tiny green shamrocks on a white background. Marked on inside: "Carrigaline Pottery Co. Ltd. in Cork, Ireland." 5" high. $20-25.

Detail of the shamrocks on the spaghetti ware bell.

Ceramic bell with pretty English cottage scene on front, made by Buckleigh Moorland Pottery, Cornish Crafts, England. $20-25.

Bone china bell from England by Coalport, white with Chinese willow pattern. *Courtesy of Tim and Kim Allen.* 4" high. $20-25.

Earthtoned pottery bell from Mexico with flower motif around central part of the base. 3.5" high. $10-12.

Glazed ceramic Clonmel Pottery bell from the island of Jamaica, decorated with three mottled bands of brown and gray. 5.5" high. $20-25.

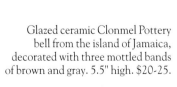

Enamel

An ancient art form, enamel on metal employs a number of painstaking but ultimately beautiful techniques, all derived from the same basic process. This process involves heating a paste made of powdered glass and water to a high temperature, then fusing it onto a metal surface such as copper, bronze, silver, or gold. The colored enamel may be applied to the surface like an oil painting, or it may be placed within small "compartments" that create a framework of intricate designs. When the compartments are formed by thin wires soldered to the metal's surface, the technique is known as cloisonné. When the compartments are formed by scratching or etching the surface, leaving small hollows or depressions behind, the technique is known as champlevé. Other techniques used in enameling include plique-á-jour (in which the metal base is removed after firing), repoussé (in which the design is embossed from underneath), and basse-taille (in which the metal is engraved or hammered to various depths).

All of these techniques produce a rich looking, often brilliantly colored effect, one with tactile as well as visual appeal. Early civilizations sometimes used enamel decoration in place of precious gems on their jewelry, so elegant was its appearance. Although China is the source of many enameled bells found in collections—and the Chinese excelled at this form of artwork—enameling has been widely enjoyed by Western as well as Eastern cultures and is used extensively on decorative objects both secular and religious.

Three small cloisonné bells from China, all with floral designs and slender cylindrical handles. Heights, from left: 3.5", 4.5", 3". $15-25 each.

Left and above:
Royal blue enamel bell from
China, with tall cylindrical
handle. This old bell is of very
thin metal and has a handmade
porcelain clapper. 5" high. $60-75.

Chinese enameled bell
with fish handle, floral
motif on base. 4.25"
high. $70-75.

Chinese enameled bell with rooster handle,
blue decorated base. 4.5" high. $75-95.

Mandarin hat button bells, survivors of a ranking system used during China's Ch'ing Dynasty, are among the most well known and popular enameled bells. The Mandarins were high ranking military or civil officials belonging to one of nine ranks. Governors and generals comprised the highest ranks, lieutenant governors and judges the middle ranks, and miscellaneous minor officers the lower ranks. To identify their rank, the Mandarins wore precious jewels of different colors on the tops of their hats, later replaced by semi-precious stones but still made of different colors. Around 1900, these stones were made into the handles of little enameled bells, two of which are shown here. From one to nine (highest to lowest), the colors of the nine official ranks were as follows: ruby, coral, sapphire, lapis lazuli or blue opaque, crystal, moonstone or white opaque, plain gold, engraved gold, and silver.

Unusual bell from China with green enameled hexagonal base. The handle appears to be made of quartz. 4" high. $75-95.

Mandarin hat button bell from China, turquoise enamel on copper base, c. 1900. The blue handle represents the third, or sapphire rank. 3.5" high. $135-150.

A second hat button bell from China, yellow enamel on copper base with robin's egg blue handle. 3.5" high. $145-160.

Interior of the yellow hat button bell, showing the pink bead clapper and the word "China" stamped on the upper section of the rim.

Red and green cloisonné bell with small loop handle, purchased in the Chinatown section of Sacramento, California.. Blue flowers decorate the front and back, green semicircles encircle the rim. 3" high. $20-25.

Although handles made of wood are common on bells, entire bells made of wood are not. While the bell's outward appearance may well be pleasing, its ability to serve as a bell (in other words, to ring) is usually hampered by the material it's comprised of. Some bells tackle the problem by incorporating a metal insert in the base, against which a flat metal clapper can be swung. Others, like most of those shown here, rely on their ancillary qualities—smooth contours or earthy colors, for instance—to win the affection of their owners.

Nicely grained wooden bell with handle reminiscent of a clothespin, originally from New Zealand. 3.5" high. $10-15.

Hand lathed wooden bell made by James Techenor, of Hoffmeister, New York. Perhaps the artist's Adirondack Mountain location inspired this bell's graceful and interesting shape. 8.5" high. $10-15.

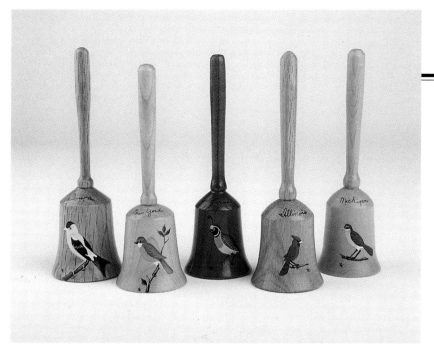

Left and center left:
Set of five wooden bells representing state flowers, birds, and types of wood, shown front and reverse. All were made in Carradus, Iowa and labeled "Carradus Originals." From left: Iowa, made from red oak with Eastern goldfinch and wild rose designs; New York, made from maple with bluebird and rose designs; California, made from redwood with California valley quail and golden poppy designs; Illinois, made from white oak with cardinal and violet designs; Michigan, made of pine with robin and apple blossom designs. All 5.25" high. $15-20 each.

Left and above:
Although wooden bells can be attractive and decorative, they typically lack the ability to produce much sound. The metal insert in the base of this traditionally shaped wooden bell is designed to help the flat metal clapper create more of a "ring." 6" high. $10-12.

Water buffalo bell from Hawaii, with twine handle and long cylindrical clapper. The bell is made of koa wood, which comes from a tree found in Hawaii. 2.75" high, 6.5" long at base. $10-12.

Mahogany bell with handle in the shape of a treble clef. Made by Homer B. Seibert, Harrisburg, Pennsylvania. 8" high. $10-15. *Courtesy of M. "Penny" Wright.*

Old wooden bell from Costa Rica, reported to be over 110 years old. It is made of cocobolo wood, a wood that is also known as golden wood and is now extinct. The bell has a wooden clapper attached by a long chain to metal loops at both ends. The upper loop pulls out of a hole in the bell's top, most likely for hanging. Tag inside reads: "El Condor / Heco en Costa Rica / Marca ® Egistrada." 7" high. $20-25.

These two miniature wooden bells by Carradus in Iowa also serve as thimble holders. Each is decorated for the holidays with a Christmas tree and holly leaves. Both 2.5" high. $5-8 each.

Mixed Materials

When it comes to their components, these bells don't fall neatly into just one category. Most have a handle and base comprised of different materials, some have an interior that belies their external appearance. Rather than detract from their appeal, however, these contrasting elements provide the bells with added interest and allure.

Possibly from England, this tall bell has a brass base decorated with enamel overlay. The base is paired with a rather sturdy looking wooden handle topped by a another small piece of brass. 8.25" high. $25-30.

Brass, tulip shaped bell with painted wooden handle and brass "crown" at very top. 7" high. $15-20. *Courtesy of M. "Penny" Wright.*

Tall, cut glass bell in green with wooden handle, purchased in Brazil. 9.5" high. $20-30.

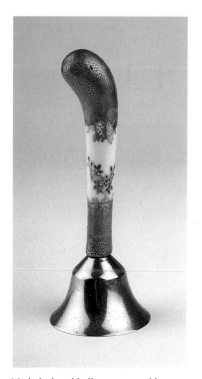

Nickel-plated bell surmounted by an unusually shaped, hand painted bone china handle from Sheffield, England. 4.75" high. $30-40.

Silver dinner bell with cherry wood handle, by Gorham. Embossed designs grace the front and back of this tall bell, which was purchased new in 1973. 8" high. $20-25.

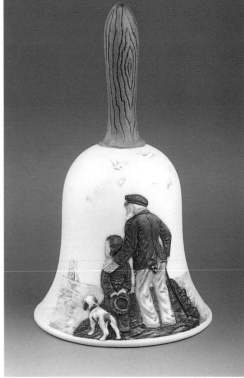

Above and right: Swedish figurine bell made of both metal and wood. As shown in the interior view, the figurine's skirt and the clapper underneath are metal. From the waist up, however, the bell is made of wood. 4.25" high. $35-40.

Porcelain bell with wooden handle, by River Shore, Ltd. Part of a limited edition Norman Rockwell series, this 1981 bell is titled "Looking Out to Sea." 6.25" high. $35-40.

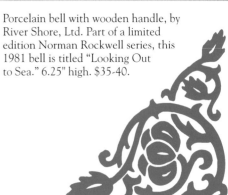

Although their individual qualities are quite dissimilar, glass and metal used in combination on a bell balance each other nicely, as illustrated on this glass bell with an oval shaped, ornately decorated metal handle. The clapper is also made of metal. 6" high. $20-25. *Courtesy of M. "Penny" Wright.*

A finely detailed French poodle made of glass sits atop this small brass bell. His front paws are upraised in a hopeful position. 4" high. $10-15.

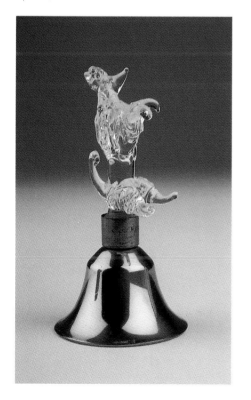

Crystal accents lend a glittery touch to this Capodimonte porcelain bell with Swarovski crystal handle and clapper. Additional crystal pieces embellish the base. 3.5" high. $55-65.

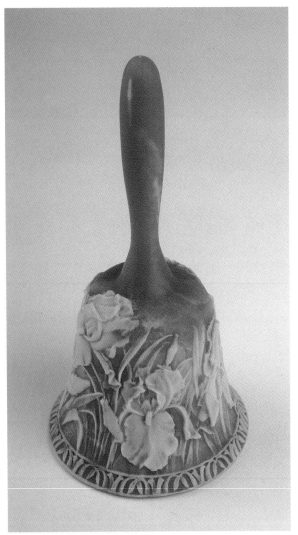

Left and above:
This Incolay® Stone bell from San Francisco is actually a "bell within a bell." The outer bell is made of variegated pink stone (a combination of quartz based minerals) with flowing white irises decorating the base. Inside rests a smaller brass bell, which in turn houses the doughnut shaped metal clapper. 7" high. $35-40.

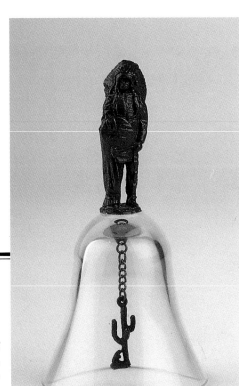

Crystal bell with figural metal handle of Native American in traditional dress, miniature cactus as clapper. 4" high. $15-20.

Chapter Three

Purposeful Bells

Although "retired" now, each of these bells was once gainfully employed—to use an analogy familiar to most of us! Their primary purpose was to serve in some capacity, although the locale and nature of that service varied widely. Some stood solemnly in church, others on the desks of now defunct restaurants or hotels. Some kept predators away from animals, other kept fussy children calmed and entertained. The sounds they produced, whether ring, chime, buzz, or clack, were their most important trait; today they are appreciated for their intriguing and often charming tales from the past.

Religious and Inspirational

For those whose faith is an integral part of their lives, the sound of bells can be an inspiring one, evoking memories of ardent messages intoned by long-robed clergymen or of peaceful reflection in a house of worship. Bells have been a part of religious observance for many centuries, joining with music, art, and ritual to augment and enhance the many forms of ecclesiastic ceremony. Their use embraces all cultures and creeds, their tones range from solemn and onerous to light and melodious. As religious accompaniments, bells have called congregants to prayer, celebrated weddings and other joyous occasions, and tolled reverently for the dead. Besides contributing to actual religious service, bells can be used to honor or memorialize historic religious figures or events.

Very old and heavy altar bell from France, bronze with original clapper. The ornately embossed base features lions' heads and turtles in the design, the handle has a tiny fleur-de-lis engraved on both the front and reverse. 5.5" high. $75-90.

Set of three altar bells with fleur-de-lis handle, made of heavy brass. 5" high, 5.5" wide. $35-50.

Heavy brass Evangelist bell with black ebony handle, openwork design on the base. The names of the four evangelists, Matthew, Mark, Luke, and John, are inscribed in Latin along a narrow band that bisects the base. Many versions of the Evangelist bell, differing in size, handle design, and complexity, can be found. 5.5" high. $50-60.

Left: heavy brass bell from Tibet with the head of Sherchin, feminine divinity representing wisdom, on the handle. 6" high. $15-20. Right: lamasery bell from Nepal, used in Buddhist religious rituals. This original bell has a beautiful tone. 7.5" high. $100-125.

Set of four altar bells with a rich patina, very old. These bells were originally used in a Buffalo, New York cathedral. 4" high. $75-85.

Figural silver bell from the Lincoln Mint, 1978. Plain silver base surmounted by figure of Jesus on the Cross. 4.5" high. $25-35. *Courtesy of M. "Penny" Wright.*

Silver bell with figural handle depicting Our Lady of Lourdes and the name "Lourdes" etched on the base. Purchased in 1962 at the Lourdes Shrine in France. 2.5" high. $15-20.

Evangelist bell mounted on sturdy wooden frame with handle for ringing, reputed to be a copy of an original that was part of a collection belonging to the Archbishop of Rheims. This unusual bell may have been used in a monastery. Overall height: 11". Bell: 5.5" high. $75-100.

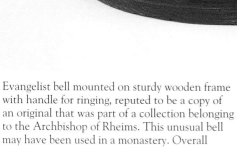

Animal Bells

From cattle to cats, four-legged creatures have always worn bells, bedecked by their two-legged caretakers with all manner of tuneful trappings intended to keep them safe, track their whereabouts, and beautify their appearance.

In some respects, animal bells represent an interesting study in the divergent attitudes exhibited towards our fellow inhabitants of the earth. Some are purely utilitarian, constructed crudely or carelessly with rough or even second-hand materials. Others combine the requisite functional nature with an aesthetic one, using decoration, color, or an interesting shape to make the bell as attractive as it is practical. One might expect the former type to be worn by animals maintained purely as commodities, while the latter, more pleasing, type might be found on animals appreciated for their many other qualities as well.

Philosophical arguments aside, animal bells possess an unmistakable ability to conjure up colorful and often vivid images: peaceful green pastures, bellowing cows, high-stepping horses, swaying elephants. Sleigh bells, those ever-popular little crotals, are among the most evocative. A group of Canadian bell collectors proves the point as they reminisce about their early memories of these fondly nick-named "jingle bells:"

Sleigh bells rang through the long tedious winters, adding a touch of cheerfulness to a dreary season. We wouldn't wish to return to those years, but all of us have memories of sleigh rides . . . the smell of fresh hay and horses, their breath white in the frosty air, laughing with friends as the bells tinkled around us, then home for hot chocolate . . . The music of sleigh bells is blanketed in sentiment. They fascinate us all: their charming tinkling, the memories they invoke, the Christmas card-like winter scenes, the music of an era past. (*The Bell Tower* 54 no. 6, 26)

Three small metal cow or animal bells. The rectangular bell on the left is 4" high; the center bell is a petite 1.75" high, marked "Zermatt" ("Switzerland"); and the bell on the right is 2.75" high and has faint, oblique lines etched on the front. From left: $20-25; $5-8; $5-10. *Courtesy of M. "Penny" Wright.*

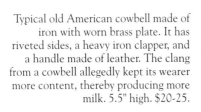

Typical old American cowbell made of iron with worn brass plate. It has riveted sides, a heavy iron clapper, and a handle made of leather. The clang from a cowbell allegedly kept its wearer more content, thereby producing more milk. 5.5" high. $20-25.

Traditionally shaped cowbell, appears to be made of copper. 3.75" high. $5-8. *Courtesy of M. "Penny" Wright.*

Pair of crudely formed animal bells, possibly made from copper. The bell on the left has a wooden clapper, its counterpart on the right has metal clapper and a narrower shape. Both 3.625" high. $8-10 each.

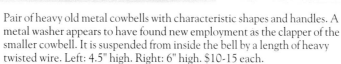

Pair of heavy old metal cowbells with characteristic shapes and handles. A metal washer appears to have found new employment as the clapper of the smaller cowbell. It is suspended from inside the bell by a length of heavy twisted wire. Left: 4.5" high. Right: 6" high. $10-15 each.

Left:
This brass cowbell has been reincarnated into an advertising tool for an early twentieth century firm. Sticker on the front read: "Compliments of "Belgard for Safeguard" / Chicago 1923." 2.5" high. $20-30.

Right:
This simple wooden cattle bell from Thailand is comprised of a central wooden cylinder flanked by two hinged clappers on either side. All three sections are suspended together from a long piece of rope. 6" high. $20-25.

Bottom two photos:
Very rare set of horse swinger bells from Persia, c. seventeenth or eighteenth century. The bells hang from two metal plates, four from the upper plate and six from the lower one; the plates are the same size but have different designs. The bells are mounted on an old wooden bedpost, not original to set. Horse swingers were used for ornamental or decorative purposes on horses. 9.5" high. $50-75.

Very rare strap of sleigh bells manufactured by Hiram and Hubbard Barton, the sons of William Barton. The elder Barton was a highly renowned sleigh bell manufacturer from East Hampton, Connecticut, who pioneered new techniques for making sleigh bells and taught his skills to a host of others in the mid-nineteenth century. This original strap with canvas backing has a double row of twenty-eight small bells along with four single graduated bells in two sizes. The twenty-eight small bells and the two smaller graduated bells have the initials "HB" engraved on the metal. One of the larger graduated bells is engraved with the initials "NS," indicating that it was manufactured by the Niles and Strong Company, another Connecticut bell manufacturer. $250-300.

Right: Detail of sleigh bell with "HB" initials. While this strap has the traditional globular shaped bells, sleigh bells were made in many other shapes. They include raspberry, strawberry, tulip, the highly prized acorn, and square.

Set of three highly burnished brass hame bells on metal bar, unusual because the frame curves upwards with a bell on each side. Hame bells are named for the part of the horse's harness to which they were attached. Center bell: 3" high. Outer bells: 2.5" high. $45-50.

Old animal bell from Austria with very faded cross design on the front. Such bells were attached by means of the swiveled hanger at the top. 4.5" high. $30-35.

Very heavy camel bell of metal, consisting of a large outer bell with handle and a smaller bell inside serving as the clapper. The outer bell has a diamond shaped emblem on each side and a heavy link chain attached to the handle. 7.5" high (excluding chain). $35-45.

Set of old metal camel chimes from Persia, originally owned by S.S. Sarna. The set consists of five open mouth bells of graduated sizes, all hung together. 13" high. $60-70.

Hand carved wooden camel bell from Kenya with two long cylindrical clappers. 8.75" from end of clapper to top. $20-25.

Elephant bell with trumpeting elephant handle, made in China. Very heavy and unusual. 4.5" high. $30-35.

Cast brass reindeer bell from Norway. The raised image on the front depicts a reindeer with characteristic antlers. 2.75" high. $15-20.

A trio of elephant bells from India showing the diversity of sizes. The characteristic pronged opening at the bottom is known as a "tiger claw;" one of the original functions of these bells was to keep tigers from preying on elephants. From left: 4.5" high, $30-35; 3.25" high, $20-25; 1.75" high, $3-5.

Handled elephant bell with coordinating design on bell and handle. 6.5" long. $10-15.

Left and below:
Elephant bell from India with matching base, both decorated in red. Bell and stand together: 4" high. $30-35.

Very heavy and unusual ornamental piece, perhaps worn as decoration by an elephant. A small elephant bell is suspended below the face of a god or goddess, while two crotal bells and two rectangular pendants hang from the ends of four long metal chains. One of the pendants has a bar on the reverse. Although its origin is unknown, it appears to be from India or perhaps Thailand. 22" total length. $60-75.

While the persistent ringing of any bell is likely to gain attention, mechanical bells are those most typically used when the bell's primary function is that of summoning assistance. Perhaps the sound made by mechanical bells—that distinctive "ding"—or the means of creating it appeal more to those seeking to announce their arrival or secure some kind of aid. After all, the action of ringing such a bell can easily escalate from a soft courteous tap to a loud impatient clamor, depending on the ringer's perception of how long he or she has had to wait!

Call bells have a long and still current history of being used in hotels, restaurants, schools, shops, banks, even private homes—virtually anyplace where one person needs to attract the attention of another. Those from the mid-nineteenth century are especially favored by collectors, as they often combined a utilitarian function with an elegant and quite aesthetic design. Many had elaborate, ornate bases topped by lustrous round canopies, the ringing mechanism located either at the top or on the side. Some of the bases were footed, others made of wood or marble that contrasted nicely with the shiny metal bell. An unusual, but rather pleasing, combination can be seen in the metal twist bell paired with a luminous, carved seashell.

Call bells were most commonly made from nickel over brass, with silver popular as well. Fancy silver serving pieces with call bells attached were often found in upper echelon homes, where they were used to summon household servants to the dining room. A selection of these highly sought pieces can be seen in Chapter Six.

Metal tap bell with decorative base comprised of leaves standing on end. Depressing the white ceramic tapper on the top moves the central clapper to the side, where it strikes the rim of the bell to produce a sound. 5" high. $60-70.

Silver hotel tap bell with twist ringer and elegant design around base. Beautiful tone and mint condition. 3.5" high. $70-80.

Right:
Brass tap bell from India with base comprised of four graduated steps. Depressing the flat lever on the left side of the bell causes the attached tapper to rise up and strike the gong type bell. 5.5" high. $8-10.

Antique, silver-plated tap bell with unusual design. Two separate hemispherical cups are joined together to form a complete ball, which is set on a marble base. Patent dates (for the finial piece only) of August 25, 1863 and April 8, 1856. 6" high. $75-80.

Old silver tap bell in the shape of a classic Grecian urn, set on a black base. Patent dates in two places on the bell read Aug. 25, 1863 and April 8 -56. 8" high. $75-80.

Metal tap bell with enameled bands of orange and yellow floral designs separated by narrower bands of light green vertical bars. Most likely of Indian origin. 5.5" high, 4" dia. $40-45.

Old bronze tap bell with side clapper, pointed finial, and pointed leaves encircling base, c. 1860. Originally purchased in Ireland, the bell is rung by depressing the lever on the side. 5" high. $50-60.

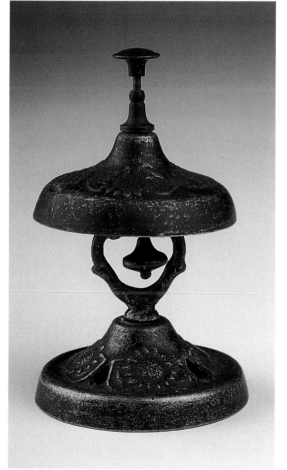

Victorian, silver-plated tap bell with egg shaped frame around bell. Patent dates on the clapper are July 25, 1875, Feb. 5, April 5, 1874. 6" high. $75-85.

Metal tap bell with raised flower design on both the canopy and the base, appears old. Depressing the plunger on top raises the bud shaped clapper to the side of the bell to make it ring. 5.25" high. $45-50.

Embossed silver tap bell with a ring that sounds like a doorbell. Although this elegant bell has no marks, the dealer from whom it was purchased stated that it originally came from the Statler Hotel in Buffalo, New York, an establishment no longer in business. 3.5" high. $60-70.

Twist bell attached to a shell which has been carved with a delicate floral design. The bell is rung by twisting the knob on top. 4" high, 5.5" wide. $35-40.

Nickel-plated twist bell with redware base, decorated with marbleized glass. "The Popular Bell" is written on top of the twister as well as on the redware bottom. 2.75" high. $50-60.

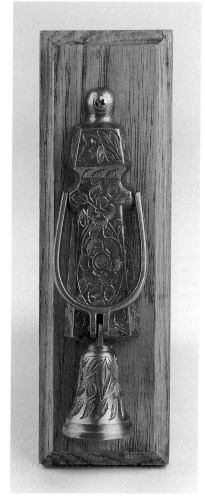

Right:
Sarna doorbell mounted on wooden base (not original to bell). Unlike most doorbells, this is a non-mechanical model; it relies on the ringing of a traditional open mouth bell to announce the presence of a visitor. 6" high (excluding base). $15-20.

Silver tap bell with screw turn top and ornate, three-footed brass base. 4" high. $60-70.

Twist doorbell made of burnished brass. The handle for this bell is attached to a brass plaque in the shape of an eagle with outstretched wings. 4" dia. $35-40.

Old metal doorbell with intricate design, mounted on a wooden base. Pulling on the string attached to the bottom of the bell activates the ringing mechanism inside. 5.5" dia. $35-40.

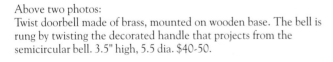

Above two photos:
Twist doorbell made of brass, mounted on wooden base. The bell is rung by twisting the decorated handle that projects from the semicircular bell. 3.5" high, 5.5 dia. $40-50.

Stroll down the aisles of any present-day toy store and you are likely to find a wide assortment of children's merchandise containing a bell or two. Bring a child along, and he or she is likely to gravitate to those very same toys, charmed by the ring, chime, or buzz emanating from inside!

As a means of entertainment for children, rattles, bells, and bell toys are as treasured today as in centuries past. From the baby waving a rattle to the toddler pulling a windup bell toy, little ones squeal with delight at the sound these actions create. As adults, we still relate to the pleasure derived from toys—that, perhaps, accounts for the attraction we feel for the playthings dating from yesteryear.

Mother-of-pearl teething rings with sterling silver bells attached might seem a bit pretentious for today's children, but they were popular in the early part of the twentieth century and sold into the 1950s. Some were decorated with scenes from children's stories or nursery rhymes, others were inscribed with the baby's date of birth and weight. Those made in the 1920s and 1930s were open at the bottom, those made later were closed. Still others used small silver crotal bells at the ends to amuse or distract a fussy baby.

Selection of old Victorian teething ring rattles. The clown figure in the center is not marked, the others are inscribed (from left): "Baby," "Mr. Bear," "Jane 7-4-28," and "The Cat and the Fiddle." Clown: $35-40. All others: $45-50.

Two sterling silver and mother-of-pearl baby rattles, c. late nineteenth century. Top: sterling rattle combined with working whistle, crotal bells attached to either side above mother-of-pearl handle, well used. 3" long. $60-75. Bottom: lollipop shaped sterling baby rattle with "hammered" appearance, mother-of-pearl handle. 3.25" long. $55-60.

Above and above right:
Rare silver teething ring with decorated, sterling silver crotal bell attached by a small loop. The front of the teething ring is decorated with an engraved floral design, amethyst gemstone, and the word "November," while the reverse reads "Santos Santiago Rubira" at the top and "1902" at the bottom. 2.875" dia. $50-60.

Selection of old celluloid baby rattles with diverse styles. From left: celluloid ram rattle attached to oval ring, rare. 4.75" high x 6" wide. $60-70; figure of little girl attached to coiled handle with loop at end, very unusual, said to be from Germany, c. 1935. 7" long. $40-50; undecorated round baby rattle with twisted green handle. 5.75" long. $35-40.

Two baby rattles made of sterling silver and mother-of-pearl. Left: three crotal bells attached to a spoon shaped mother-of-pearl handle by a small ring, no marks, probably c.1900. Right: combination baby rattle and teething ring, with space to engrave the child's date of birth, weight, and time of day, marked "SLB Sterling." $55-65 each.

Old silver baby rattle embellished with beaded trim, inscribed "Anne" on the ivory handle. No other marks. 5" long. $100-115.

Toys that generate ringing bell sounds when pushed or pulled along the floor are enduring favorites with little ones. Metal toys called bell ringers first came into vogue during the late nineteenth century. They combined wheels with bells and were widely manufactured from about 1880 to 1920. While metal may be more durable, other materials have been used for toys employing the same concept. Several examples in wood or celluloid are shown here; each incorporates a bell that is activated by the child pulling, pushing, or twisting some part of the toy.

Old bell pull toy consisting of a painted wooden dog attached to two metal wheels with bell centered between them. Pulling the dog by the string in front causes the bell to ring. 12" long. $100-125.

Detail of the pull toy, showing the bell between the two red wheels.

Left and below:
Very old bell pull toy, c. 1900. The toy consists of four metal wheels joined by two long metal strips. Each of the two "axles" has a metal bell centered between multi-colored wooden balls arranged like spokes. Pushing or pulling the toy rings the bells. 9" long. $120-130.

Below:
Celluloid and tin windup bell toy made in Occupied Japan, c. 1940s-1950s. Two ducks and a chick ride in a green sleigh or wagon pulled by a perky bunny. Wheeling the sleigh causes the bell mounted on the back to ring. 7.5" long, 4" high. $125-135.

Below right:
Another celluloid and tin windup toy, this one of a boy riding a red and green scooter with bell mounted in the back. Twisting the windup mechanism rings the bell. 5" high, 3.5" long. $125-135.

Chapter Four

Beauteous Bells

While the bells in the previous chapter all had a "job" to do, those shown here have a role as well (albeit one based more on appearance than sound): they are designed to bring enjoyment and beauty to the surroundings of their owners, a role they fulfill most affably. Here you will find modest but cherished souvenir bells, convention bells with geographical themes, regally dressed figurines, whimsical holiday ornaments, diminutive bell jewelry, and more. They vary in size, shape, style, and monetary value, yet each exemplifies with ease the unmistakable charm and charisma so typical of bells.

Souvenir Bells

It would be hard, no doubt, to find a bell enthusiast whose collection does not include at least a handful of souvenir bells. Generally low-priced and widely distributed, souvenir bells lack the notability and desirability of antique or finely crafted bells, but do have a charm all their own. Colorful reminders of places visited, they usually sport the name of a city, country, or special attraction. In some cases, of course, a well-known image from the area is all that's needed to bring back those fond memories.

Two souvenir bells from the U.S. Capitol, Washington, D.C. The bell on the right also features scenes of the Washington Monument and the White House. Both 5.5" high. $10-15 each.

Three souvenir bells from the state of Alaska . All 4" high. $5-10 each.

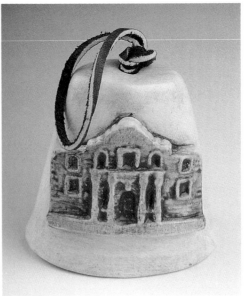

Ceramic souvenir bell from Texas, depicting the famous fort known as "The Alamo" on the front. 3.25" high. $10-15.

Souvenir bell from Mt. Rushmore illustrating the famous heads of U.S. presidents carved into the mountain. Reverse of the bell reads: "Mt. Rushmore National Memorial Shrine of Democracy." 3.25" high. $20-25.

It's not hard to guess where this bell is from—the bright red lobster with matching base is a souvenir from Maine. 4" high. $5-10.

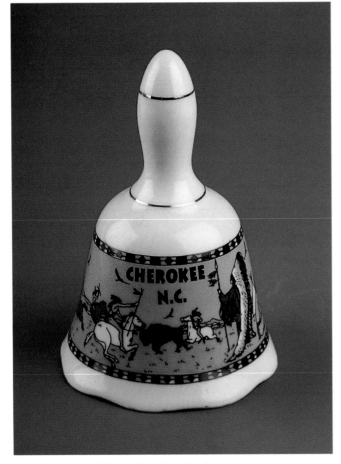

Souvenir bell from Cherokee, North Carolina, with scene of Indians hunting buffalo. 4.25" high. $10-15.

Small ceramic bell, a souvenir from the Amish country. 2" high. $5-8.

Souvenir bell from Niagara Falls, Canada. Glazed ceramic with picture of falls on base and flower design on handle. 5.25" high. $10-15.

More elegant than the typical souvenir bell, this old, cowbell shaped bell comes from Brussels and appears to be made of flint glass. The front features the frosted image of a palace, along with the words "Souvenir de Bruxelles." 4.5" high. $45-55.

Right:
Colorful souvenir bell from Wales with unusually shaped handle and shield of Wales on reverse. Marked inside: "'Yetka' Fine Bone China Great Britain." 4.25" high. $20-25.

Three souvenir cowbells with brightly colored flower decorations. The small bell on the left is only 1.75" high; it has an applied flower decal on the front along with the name "Oberstdorf," a ski resort in Germany. $5-8. The center bell features hand painted flowers and is 3" high. $10-15. The bell on the right is also hand painted and 2.5" high; on its reverse it reads: "Pilatus-Kulm / 2132m - 7000 ft," a reference to the summit of Mt. Pilatus in Switzerland. $8-10.

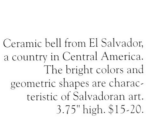

Ceramic bell from El Salvador, a country in Central America. The bright colors and geometric shapes are characteristic of Salvadoran art. 3.75" high. $15-20.

Tiny souvenir bell from the German city of Rothenburg, made of porcelain by Reutter. 1.5" high. $10-15.

A bug-eyed frog with a pink belly forms the handle of this ceramic souvenir bell from Puerto Rico. 4.75" high. $15-20.

Brass souvenir bell from Guatemala with long-tailed Quetzel bird as handle. 4" high. $15-20.

Commemorative
and Convention Bells

While they are found in a multitude of styles, commemorative bells share the common theme of paying homage to an event or institution—in some cases, even a worthy individual. Many embody patriotic or historical subjects, such as those shown here that honor America's Revolutionary War or Britain's coronation of Queen Elizabeth II. Others are designed to celebrate an important anniversary. It's not immediately apparent, for example, that the two Fenton bells on page 90 are commemoratives. They were each created, however, to commemorate the 75th and 90th anniversaries respectively of this prolific American art glass company.

Similar to figural and figurine bells (covered in the following section), commemorative bells sometimes require a bit of sleuthing on the part of a collector to determine the bell's origin and just exactly who or what is being commemorated. For those who enjoy a challenge, this can greatly increase the interest and ultimate enjoyment of a particular bell!

Brass bell with gold mesh surrounding the base, commemorating the life of Anna Warner Bailey (1758-1851), a colorful folk heroine from Groton, Connecticut who made a unique contribution to the War of 1812. The bell's handle is comprised of a central cylinder with two strands of gold chain originating at the very top and winding around the bottom; the inscription at the bottom of the base reads: "Groton / Mother Bailey 1812 / Anna Warner / 1781." Known as "Mother Bailey," Anna Warner Bailey developed a fierce love of independence—and a concomitant hate for the British—when one of her uncles was fatally wounded in 1781 during the Revolutionary War. Years later, she was the proprietor of a Groton inn when the War of 1812 broke out. Anticipating an attack by the British on New London harbor, an American soldier was sent to Groton in search of flannel— badly needed for wadding to load muskets. Most of the townspeople had already fled, but Mother Bailey stayed behind. Encountering the soldier on the street and hearing of his need, she immediately stepped out of her large flannel petticoat and grandly donated it to the cause. Although the British never attacked after all, Mother Bailey was hailed as "the war's greatest female patriot" and a chapter of the Daughters of the American Revolution (DAR) was founded in her honor. This Mother Bailey bell was manufactured by the J.E. Caldwell Company of Philadelphia, Pennsylvania; on the cylinder portion of the handle is engraved: "Caldwell patent." 3.5" high. $65-75.

Heavy brass bell from England commemorating one of the twentieth century's most important royal events: the coronation of Britain's current queen. Inscription around the base reads: "Coronation of HM Queen Elizabeth II 1953." The bell's metal comes from old melted down church bells. 3" high. $85-95.

Commemorative brass bell from the Canadian province of New Brunswick, handle in the shape of a cross. Inscribed around the middle of the base is: "Relic Cathedral Fire Fredricton, N.B." Founded in 1783, Fredricton is the capital city of New Brunswick. 3.5" high. $20-25.

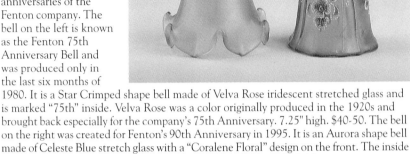

Two versions of bells commemorating Kiwanis International, a civic organization founded in 1917. On the left is a heavy brass speaker's bell with wooden mallet, marked on the bottom: "Service Club Supply House, Toronto, Canada." 8.75" high. $65-75. The bell on the right has the same Kiwanis logo on the handle but is only 3" high. $20-25.

These two glass bells were both created to commemorate special anniversaries of the Fenton company. The bell on the left is known as the Fenton 75th Anniversary Bell and was produced only in the last six months of 1980. It is a Star Crimped shape bell made of Velva Rose iridescent stretched glass and is marked "75th" inside. Velva Rose was a color originally produced in the 1920s and brought back especially for the company's 75th Anniversary. 7.25" high. $40-50. The bell on the right was created for Fenton's 90th Anniversary in 1995. It is an Aurora shape bell made of Celeste Blue stretch glass with a "Coralene Floral" design on the front. The inside is signed "handpainted by P. Hyhurst—Fenton 90th." 7" high. $35-40.

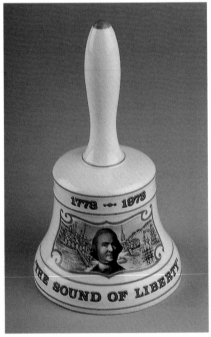

Tall porcelain bell commemorating the 200th anniversary of the Boston Tea Party, a watershed event in the American quest for freedom. The inside of the bell reads "Hammersley Fine Bone China Made in England © 1973," along with historical information describing the pictures on either side of the bell: "Bostonians cheer from the quayside as the 'Mohicans' dump the English tea over the side," and "Paul Revere's view of Boston in 1773; together with Sam Adams arch-planner of the Tea Party." Distributed by Schmid Bros., this bell was one from a series of four, all illustrating scenes from America's early history. 7" high. $20-25 each. *Courtesy of M. "Penny" Wright.*

Left:
Lead crystal, hand cut bell with swirled handle, commemorating the 1980 Winter Olympic Games held in Lake Placid, New York. Small sticker on the bell's front identifies its maker as the Cavan Co. 7.5" high. $80-85.

Series of colorful porcelain bells from the Danbury Mint commemorating the American Bicentennial, made in Germany. Pictures on both the front and reverse of each bell show Revolutionary War scenes as well as portraits of early American patriots. From left: scene of Washington crossing the Delaware River, December 25, 1776, map of Delaware River on reverse; portrait of George Washington with signature below, scene of Mount Vernon, Washington's home, on reverse; scene of Washington at Valley Forge, 1777, scene of winter at Valley Forge on reverse. All 6.5" high. $20-25 each. *Courtesy of M. "Penny" Wright.*

From left: scene of Betsy Ross sewing flag, image of flag and the words "First Official Flag of the New Nation Created by a Resolution of the Continental Congress" on reverse; scene of Paul Revere's April 18, 1775 ride, map of Revere's route and poem by Longfellow on reverse; scene of three fife and drum players with flag, "Spirit of '76" plus drums, flags, and trumpets on reverse. All 6.5" high. $20-25 each. *Courtesy of M. "Penny" Wright.*

Left: scenes on both sides of the Battle of Yorktown, October 9, 1781. Right: scene of the Battle of Bunker Hill June 17, 1775, map of battle site on reverse. Both 6.5" high. $20-25 each. *Courtesy of M. "Penny" Wright.*

Portrait of John Paul Jones with signature below, scene of ships in battle and "'I have not yet begun to fight' September 23, 1779" on reverse. 6.5" high. $20-25. *Courtesy of M. "Penny" Wright.*

From left: portrait of Thomas Jefferson with signature below, scene of Monticello, Jefferson's home, on reverse; signing the Declaration of Independence, July 4, 1776, scene of Independence Hall in Philadelphia on reverse; portrait of Benjamin Franklin with signature below, scene of Franklin flying kite on reverse. All 6.5" high. $20-25 each. Courtesy of M. "Penny" Wright.

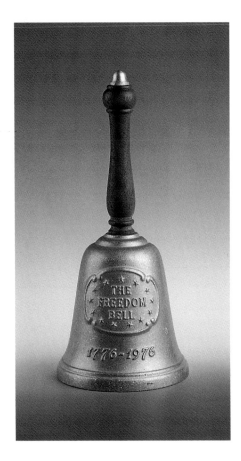

"The Freedom Bell" also commemorates the 1976 American Bicentennial. Metal bell with wooden handle has inspirational phrase extolling freedom on reverse side. 10" high. $35-40. *Courtesy of M. "Penny" Wright.*

Reverse of "The Freedom Bell."

Colorful ceramic bell commemorating the New York World's Fair, 1964-1965. The red haired little girl who serves as the handle was named "Wendy;" together with her twin brother, "Peter," she was illustrated on a host of World's Fair items, now very collectible. $40-45. *Courtesy of M. "Penny" Wright.*

This impressive looking bell sitting on its own engraved platform is known as the "General Grant Bell" and serves as the official symbol of the American Bell Association; a drawing of the bell, in fact, appears below the return address on each issue of ABA's newsletter, *The Bell Tower*. First owned by General Ulysses S. Grant, the bell was originally used in the nineteenth century to call farm workers on Grant's Missouri plantation to dinner. Later given to a bakery worker who helped fill the weekly bread order for General Grant, the bell was handed down to that worker's daughter—a bell enthusiast named Augusta Littman who served as ABA's president in 1959. As recalled by that early president, the bell became ABA's "official" bell in a rather unofficial manner. Following the organization's second annual convention in 1947, members were invited to Mrs. Littman's home for lunch. Admiring the bell collection afterwards, one member remarked offhandedly that it would be nice to have an official organizational bell. A second member suggested it be something "American," whereupon Mrs. Littman handed her the General Grant bell as a suggestion. Readily accepted, the General Grant bell has served in that official capacity ever since. The names on the base of the platform represent former presidents of ABA. Complete refurbishing of the platform and name plates was done courtesy of the ABA Past Presidents in 1995. Bell: 9" high. Platform: 2.5 high, 10" sq. Value undetermined.

Detail of platform, showing a bust of Ulysses S. Grant and the three names by which the organization has been known. Also added courtesy of the ABA Past Presidents in 1995.

Detail of the bell's handle. "N.B.C.C." stands for National Bell Collectors' Club, the original name for the organization now known as ABA.

Included here are several bells representing American Bell Association (ABA) convention bells, all limited editions. The annual ABA convention is held each year in a different location around the country and specially commissioned convention bells are made available to the enthusiastic collectors who attend. Very popular with ABA members, the convention bells are often designed by local artists and reflect the flavor of the state hosting the convention—a dolphin from Florida, a New England meeting house from Connecticut, a sailboat from San Diego.

Detail of the gavel shaped clapper from the New England Meeting House bell shown above right. Notice the initials "ABA" over the front door of the meeting house, as well as a smaller "cornerstone" version visible at the lower right side of the building.

Trio of American Bell Association (ABA) convention bells, each commemorating a specific year's convention. Left: bronze bell from the 49th ABA convention, held July 10-13, 1994 in San Diego, California. This bell was made by the Art Foundry, San Diego, in a limited edition of 300. The clapper is in the shape of a small anchor. 4" high. Center: bronze bell from the 43rd ABA convention, held June 26-29, 1988 in Hartford, Connecticut. Shaped in the image of a New England Meeting House, this bell was created by Daniel J. Riccio in a limited edition of 300. The clapper is a tiny "house gavel." 2.25" high. Right: bronze bell from the 50th ABA convention, held June 25-28, 1995 in St. Louis, Missouri. This bell was designed by members of the Gateway Arch Bells Chapter and Kurt Mager of Mager Metal Art, Ltd., Des Plaines, Illinois, in a limited edition of 300. Its profile is intended to simulate the famous Gateway Arch in St. Louis and the fleur-de-lis handle symbolizes the city's French heritage. 3.25" high. $35-45 each.

Left:
Heavy bronze commemorative bell from ABA's 1982 convention, held in Orlando, Florida. Florida's state animal, the dolphin, serves as a dramatic handle for this bell, which weighs 2.5 pounds. 6.5" high. $30-35.

Right:
Pewter bell from ABA's 1992 convention, held in Appleton, Wisconsin. A rather small base is surmounted by the figure of a town crier sculpted by artist Michael Ricker. Limited edition of 300. 8" high. $25-30.

The clapper from the 1987 convention bell is a colorful ceramic pineapple, long known as a symbol of hospitality.

Glazed ceramic bell from ABA's 1987 convention, held in the state of Hawaii. Designed by Hawaiian artist Ele in a limited edition of 400, this bell shows an ancient Hawaiian petroglyph of a dancer with rattles on one side and the traditional greeting "Aloha" on the other. The designs known as petroglyphs are found in Hawaiian caves and on the rocks surrounding the islands. 6.5" high. $20-25.

Figurals and Figurines

Highly prized by collectors, figural and figurine bells combine a pleasing countenance with an often intriguing history. While they share many characteristics (and are therefore shown together in this section), figurals and figurines differ with regard to how much of the bell is actually comprised by the figure. On figural bells, only the handle or uppermost part of the bell contains the figure, usually of a person or animal. The handle of a figural bell may show the entire body, just the upper torso, or sometimes merely the head. Figurine bells, in contrast, are those composed entirely by the figure, the bell body usually serving as the skirt or robe worn by the figure. Given that most figurine bells are feminine in nature (due to the ease with which a woman's dress can be incorporated into the bell), such bells often radiate an undeniable charm and grace derived from the fashionable and elegant outfits they "wear."

One of the most interesting aspects of figural and figurine bells is their ability to portray—often with realistic detail—the clothing, hairstyles, or mannerisms characteristic of a particular country, profession, or time period. Among the bells shown here are those depicting individuals from a variety of countries, including France, England, The Netherlands, Germany, Korea, Mexico, and Chile. Many are easily recognized by the typical dress or "look" of the bell, others require a bit of geographical research to accurately place the nationality of the figure represented. In *The Collector's Book of Bells*, Lois Springer provides a few hints on determining the regional origin of some figurine bells: *(continued on following page)*

Bronze wind bell from the 1990 ABA convention, held in San Antonio, Texas. Designed by Max Greiner, Jr. of Creative Designs Concepts Inc., Kerrville, Texas, the bell features a silhouette of Texas in the circular segment at the top. 13" long. $30-40.

Even without any personal identifications of the figures involved, those dressed in regional or period costumes offer an engaging study in terms of fashion history. French faience figurine bells display the varied native costumes of the old provinces, each with its distinctive headdress. Studying the customs that prompted each manner of dress provides an excursion into French provincialism. The headpiece on a figurine is often a convenient index to the period represented. A butterfly headdress on a woman betokens a Flemish style of the late Gothic era in the fifteenth century. The tall conical hennin marked the ultra in head fashions during medieval days, and several variations of it appear on various brass portrait bells (Springer 1972, 177-178).

Literary and historical figures are understandably popular with the artists and manufacturers of figural and figurine bells, and some well-known characters have been immortalized on more than one bell—each in a variety of poses and sizes. Historical figures are shown in the following section, but here you will find a few diverse folks from literature: Don Quixote, Tinkerbell, and Mr. Micawber from *David Copperfield*.

No discussion of figural and figurine bells would be complete without mention of the emotional impact such bells may impart. Perhaps even more than recognizable characters, the depiction of "ordinary," genre type individuals has the potential to evoke strong feelings of nostalgia, sentiment, or empathy. Gazing upon either of the bells by artist John McCombie, for example, is likely to stir up memories from one's own past, leaving a deep and lasting impression. Similarly, the figural bell from the Lincoln Mint of a stylized mother and daughter will no doubt elicit tender feelings from all but the most hard-hearted of beholders!

Brass bell with figural handle of the Lincoln Imp. He sits with knees drawn up and a rather mysterious smile on his face. According to legend, the Devil sent imps to England's Lincoln Cathedral, where they turned those who performed "devilish acts" to stone and hid the workmen's tools overnight. 4.75" high. $25-30.

Tinkerbell, the spritely fairy from *Peter Pan*, forms the handle of this small brass figural bell. 3" high. $15-20.

Pewter bell from Belgium with small figure of Pan, Greek god of flocks and shepherds, serving as the handle. 3.75" high. $30-40.

Small brass figural bell, handle in the shape of what appears to be an Asian god or goddess. 3.5" high. $10-15.

Silver-plated bell with figural handle of guardian angel, made by the Danbury Mint in 1974. The angel is styled after one from a sculpture by the artist Fellini. 4.75" high. $20-25.

A similar bell from the Danbury Mint, this one from 1975. Plain, silver-plated base surmounted by figure of angel with clasped hands. 4.5" high. $20-25. *Courtesy of M. "Penny" Wright.*

This figural brass bell with a Middle Eastern motif features the torso of a praying figure as the handle and a design of mosques encircling the base. 3.25" high. $15-20.

Right and far right:
Similar to the Danbury Mint bells above, an angel motif appears on these two silver-plated figural bells by Adolfo Procopio. One shows an archangel holding a scythe, the other features an angel in a long gown holding a trumpet. The artist's signature is engraved on the back of both bells. Both 4.5" high. $20-25 each. *Courtesy of M. "Penny" Wright.*

Figurine bell of a woman in French peasant garb. Her dress has crisscross detail on the bodice, her hat has wide bouffant sides. Marked "Belgium" in capital letters just below the waist on reverse side. 2.875" high. $30-35.

Looking as if they might break into dance at any moment, these three figurine bells are all made from brass and depict working folk from The Netherlands. Left: Dutch woman with hands on hips, wearing apron with pockets, marked "Reg'd / Made in England" on bottom of the clapper. 3.25" high. $25-30. Center: Dutch boy with hands on hips, wearing square shaped cap. 2.75" high. $15-20. Right: Dutch woman with one hand on hip, the other carrying a bucket. 2.875" high. $10-15.

Two brass figurines of Dutch women, similar in shape and style. The figurine on the left wears a Dutch cap and has her apron filled with apples. Stamped "England" on the side, it is 3" high. The figurine on the right appears to be carrying a scarf and was made in Belgium. 2.75" high. $20-25 each.

Heavy brass figurine of young girl in Dutch style cap holding an umbrella in her left hand and a basket in her right. She wears a shawl tied with a bow in the front. This same bell is advertised in an old catalog of the Pearson-Page-Jewsburg Company, Ltd., of Birmingham and London, England. It is not certain, however, if this particular bell is an original or reproduction. 4" high. $40-45.

Made of heavy brass, this figurine bell depicts a woman wearing a Dutch style cap and apron, holding the sides of her skirt while curtsying. A pair of boots serve as clappers for the bell, which was purchased new in 1975. 5" high. $35-40.

Brass figurine of a Kentish Maid from England. Legend decrees that all maidens born south of Britain's River Stour be known as Kentish Maids; those born north of the river are called Maids of Kent. 2.75" high. $18-20.

The clappers from the Dutch woman figurine.

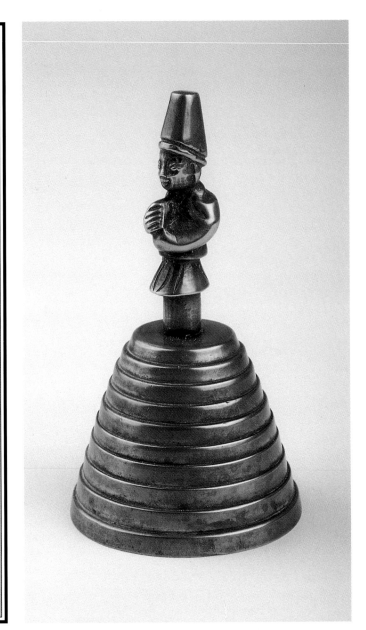

Old brass figural bell with beehive base and handle in the shape of a male figure wearing what may be a Turkish fez. 3.25" high. $15-20.

Bronze figurine of a Bavarian woman with a full, bouffant skirt, c. 1850, shown front and reverse. Notice the detail on the figurine's hair and the back of her skirt. 5.5" high. $95-100.

The stately figure atop this heavy old bronze bell appears to a native of India. He stands with arms folded and wears a toga made of feathers and a turban on his head. The bell was purchased in York, England. 6.5" high. $35-50.

Left and far left:
Figurine bell of heavy brass depicting a peasant woman from Korea, shown front and reverse. The object in her hand is thought to be either a fan for kindling the fire or a long knife. Possibly made by Korean peasants from bullet casings. 3.5" high. $25-30.

A carpenter wearing a long apron that ties in the back serves as the handle of this heavy brass figural bell, which may represent St. Joseph. The hammer in his hand moves up and down. 6" high. $45-65.

Brass bell made in Cameroon, Africa, with figural handle depicting a man holding a gourd or maraca. The base of this contemporary bell is an African hut with a straw roof. 8" high. $40-45.

Above and below:
Heavy bronze figural bell of two "Water Girls," also by John McCombie. McCombie dedicated this bell to "all the hardworking women who often go unnoticed and unrewarded for their perseverance and persistence doing menial chores." It is cast in the Lost Wax method, signed and numbered by the artist on the inside. As in "Old Friends," the bell's clapper is shaped like a realistic cowboy boot. 6.25" high. $160-170.

The mutual love of man and dog are realistically depicted in this bronze bell by John McCombie, of Indiana, Pennsylvania, aptly titled "Old Friends." Cast in the time-consuming Lost Wax method, this bell is #12 from a limited edition of 150 and features a clapper in the shape of a cowboy boot. The artist's vision for the bell grew out of the relationship with his own dog, as described in promotional material for "Old Friends": "The love of my faithful dog, Pepper, inspired me to put us in bronze. There's a bonding between us that grew over the years—we became 'old friends.'" 7" high. $160-170.

Sterling silver bell with figural handle of Don Quixote, the fictional character who fancied himself a knight in Cervantes's famous tale. Purchased in Majorca in 1964. 4" high. $35-40.

Figural brass bell, probably depicting the character of Mr. Micawber from Charles Dickens's classic novel *David Copperfield*. 3.75" high. $20-30.

Twin figurine bell, made of black iron by the Iron Art Company, Phillipsburg, New Jersey. Said to depict the wicked stepsisters from the fairy tale *Cinderella*, the figurines each have an individual clapper. 4.5" high. $20-25.

Individual clappers from the twin sisters bell above.

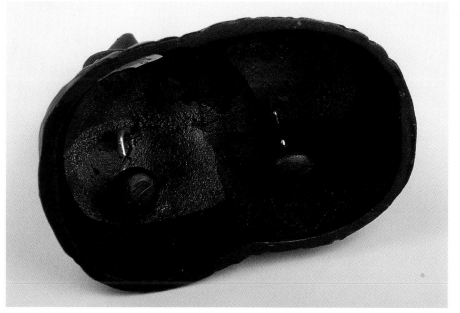

Very old and heavy brass figurine bell, depicting a woman wearing a shawl, apron, and headdress, holding what appears to be a bowl in her right hand. Very worn on sides, original clapper. This bell resembles another bell known as the Mother Hubbard bell, but is not the same. 4.75" high. $60-75, based on existing condition.

Unusual brass figurine bell with deep patina, fine detail. The aristocratic looking woman (possibly the "Goddess of Flowers") wears a low cut ballgown with ruffles around the skirt and appears to be holding a handkerchief in her right hand. Her hair is held high with a comb in the back. 4.5" high. $75-85.

Heavy old brass figurine bell of woman wearing a mob cap and carrying a long handled purse, known as the "Lucy Locket" bell. This figure is thought to be modeled after the character of Lucy Locket—who "lost her pocket"—in an old Mother Goose poem. The clappers are shaped like a pair of legs and feet. 5" high. $110-125.

Leg shaped clappers on the "Lucy Locket" bell.

Small metal figurine bell from Germany with an interesting shape. The woman wears a wide skirt accented with a shiny gold apron and bodice; her head is covered by a three cornered cap. 2.75" high. $25-30. *Courtesy of M. "Penny" Wright.*

Silver-plated figural bell by Reed & Barton, woman holding songbook forms handle. The artist's signature, Adolfo Procopio, is engraved on the back. 4.5" high. $35-40. *Courtesy of M. "Penny" Wright.*

Right:
Figural silver bell from the Lincoln Mint, 1978. Plain base surmounted by figures of a mother and daughter in loving embrace. 4.5" high. $30-35. *Courtesy of M. "Penny" Wright.*

Silver-plated figural bell by Reed & Barton, also by Procopio. A bearded man in pilgrim garb with cloak, hat, and rifle forms the handle. 4.5" high. $35-40. *Courtesy of M. "Penny" Wright.*

Figural bell from the Lincoln Mint, 1978. Plain silver base surmounted by figures of woman and man singing in front of lamp post. 4.5" high. $30-35. *Courtesy of M. "Penny" Wright.*

Right:
Figural silver bell from the Lincoln Mint, 1978. Plain silver base surmounted by figure of Indian holding corn with baby in papoose on back. 4.5" high. $30-35. *Courtesy of M. "Penny" Wright.*

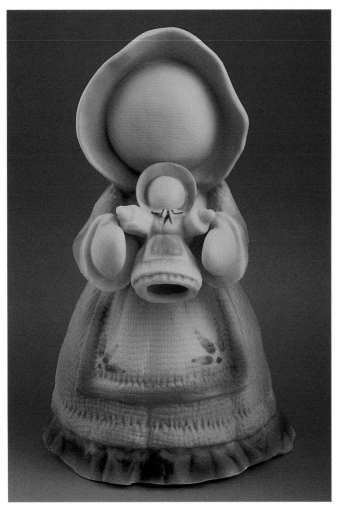

This bisque porcelain figurine of an Amish mother and her baby combines a bell with a thimble holder. The thimble holder "baby" wears a miniature version of her mother's outfit and rests between the mother's outstretched arms. The baby is not attached and can be removed for use as a thimble holder. Mother: 6.5" high. Baby: 2" high. $20-25.

Tall, hand painted ceramic figurine bell from Mexico, depicting a woman in native costume holding a brown earthenware bowl. 6.5" high. $20-25.

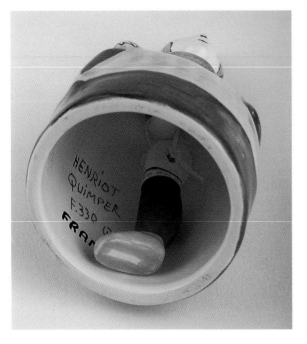

Left and far left:
Glazed faïence figurine bell from Quimper, France, of woman carrying a brown purse and umbrella, matching clapper shaped like a leg and foot. Marked inside "Henriot Quimper / F. 330 / QM / France." "QM" most likely represents the artist's initials for this hand painted bell. 5" high. $55-65.

Angel figurine bell with wings and clasped hands, holder for candle behind wings. This bell comes from Oxaca, Mexico and is made of lava rock. 4.5" high. $10-20. *Courtesy of M. "Penny" Wright.*

Glazed ceramic figurine bell depicting a long-nosed witch holding her broom, green kerchief tied under her chin. 4.5" high. $10-15.

Ceramic figurine bell, a souvenir from the country of Chile. 3" high. $15-20.

Two similarly styled ceramic figurine bells in bright colors. The bell on the left is hand made and hand painted by the Alfaraz Workshop in Madrid, Spain. 5.5" high. $20-30. Her smiling companion on the right is from Portugal. 5" high. $20-30.

Pair of glazed porcelain gnome bells with short sqatty statures and tall pointy hats, a "Mother" and "Bride." Made in Japan for the Gorham Co. 6" high. $15-20 each.

Bisque porcelain figurine bell of woman in fancy dress with upswept hair, made in Japan. The design and color of the woman's costume are characteristic of the French court in the seventeenth and eighteenth centuries. 4.25" high. $20-25.

Boot shaped clappers on the bisque figurine.

Bisque porcelain figurine bell with pink, boot shaped clappers under a tiered skirt, hands held demurely below a trio of three roses at the waist. Most likely made in Japan. 5.25" high. $20-25.

Porcelain figurine bell of young woman in fancy flowered dress. The red rose held in her hands matches the flowers on her pale green dress. 4.5" high. $25-30.

This hand painted bisque porcelain figurine bell representing "December" is part of a series featuring months of the year and their special flowers. It is marked inside: "December / Angelica with Poinsettias / ©1986 Enesco Imports Corp / Made in Mexico." 5" high. $20-25.

Porcelain figurine of young girl with contemplative look and hands folded, by Lladro, of Valencia, Spain. Known as the "Communion Bell," it is mostly white with delicate pink accents and Lladro markings inside. 4.75" high. $100-120.

Pair of glazed ceramic figurines, hand-crafted by Audrey Pollard of Streetsville, Ontario, Canada. The woman on the left wears a wide brimmed hat with matching cape and carries a basket of dried flowers. Her companion on the right has the same demure expression but is dressed in traditional nurse's garb. Both 7.5" high. $25-30 each.

Whimsical ceramic figurine of a stylishly dressed cat, made by artist Janice Joplin of San Antonio, Texas. Notice the fish shaped purse held in the cat's right hand—er, paw! 10" high. $35-40.

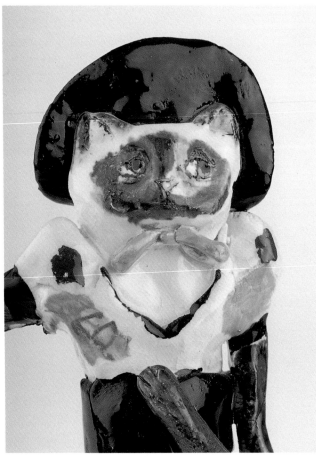

Detail of the cat figurine's expressive face.

Left:
Two headed festival bell from Mexico, showing the kind of two faced masks used in local celebrations. One side has the face of a bow-tied owl, the other a human face surrounded by what appears to be white feathers. Made of local clay and baked underground. 7.5" high. $15-20.

Below:
Detail of the two heads on the festival bell.

Glass figurine bell, "Louise," made by the Boyd Art Glass Company of Cambridge, Ohio. The woman's bonnet and flounced skirt are blue, her head and torso have a paler, almost white coloration. A tag on the side reads: "3rd annual bell 1981." 4.5" high. $25-30.

Some are more famous (or infamous) than others, but all of the individuals shown here have left their mark on history in one form or another. Depictions of noteworthy people are common on bells (particularly the figural and figurine types) and those shown here are but a small sampling of the multitude available.

More than one image of an individual is sometimes found on bells, just as many paintings or photographs may be made of the same person over the course of his or her lifetime. In addition to the Jenny Lind bell with the ruffled skirt, for example, another, less common version shows the celebrated singer holding up both sides of her layered, bouffant skirt (similar to the pose seen on Marie Antoinette). The upraised skirt causes the lower edge of the bell to be curved, hence it is sometimes known as the "Rocking Jenny Lind."

Studying the history of people shown on bells can be fascinating, but the identity of some eludes even the most determined or persistent collector. This shouldn't be considered a failure, however. As collectors Blanche and Stanley Kleven note, "[i]t does not detract from the value and the beauty of a bell if we cannot name the character. It gives us pleasure to be able to do so, but we can learn so much by simply identifying a time in history or a class in society." (Kleven 1993, 24)

Brass or bronze figurine bell of Queen Elizabeth I wearing a "ruff" at her neck and holding a fan with both hands. Purchased in Tours, France, this bell has a nice detail and a blue/green patina. 4" high. $65-75.

Heavy old brass figurine bell, also of Queen Elizabeth I, finely detailed. She holds one hand at her waist and wears a dress with a high collar dress and shoulders. 5.25" high. $50-60.

Small brass figurine bell of Marie Antoinette, made in Belgium. Notice the fine detail, especially in the fabric of the elaborately designed gown. French registry mark found on the back at the base of the skirt: "102 Deposé." 3.5" high. $50-75.

Known as the "Madame Pompadour" bell, this very heavy, finely detailed brass figurine bell depicts an upper class French woman in fancy gown and headdress. She is said to represent Madame Pompadour, the eighteenth century mistress of Louis XV. 6.25" high. $65-75.

Heavy old figurine bell of Jenny Lind, known as "The Swedish Nightingale." Born in Stockholm in 1820, Lind was widely acclaimed for her diverse singing talent. Here she wears a long tiered dress and a poke bonnet tied under the chin. Her left hand is turned at the waist and rests on the back of her dress. Although purchased in Belgium in the early 1900s, this bell is probably of British origin, as it is advertised in an old catalog of the Pearson-Page-Jewsburg Company, Ltd., of Birmingham and London, England. Legs and feet comprise the bell's clappers. 5" high. $95-100.

The clappers of the Jenny Lind bell.

Brass figurine of Jenny Jones, sometimes known as Mary Jones. This legendary young girl came from Cardiganshire, on the west coast of Wales. Intent on purchasing a Bible, Mary is said to have saved her money for six years and then walked barefoot around 1802 to the town of Bala, some twenty-five miles away. From this story came the founding of a Bible society in England and later the American Bible Society. 4.25" high. $20-25.

The image (as well as the name) of Jenny Jones also appears on this figural bell purchased at a flea market in England. It is a typical English bell with decorated base and flat, one-sided handle. 6.75" high. $35-40.

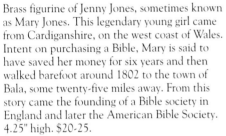

Brass figurine of Sally Bassett, a slave from Bermuda said to have been burned at the stake in 1730 for attempting to poison her master's family. This bell, which portrays Sally resting on a cane, was purchased in Bermuda in 1960. It has an English registry mark on one side. 4" high. $30-40.

A smaller version of Sally Bassett. 2.5" high. $20-25. *Courtesy of M. "Penny" Wright.*

Figural bell from California with handle representing Fra. Junipero Serra, the eighteenth century Franciscan priest considered the founding force behind the network of California missions. Inscribed around the rim: "Fra Junipero Serra Pres California Missions." 3.75" high. $25-30.

Farmyard to Forest—Figural Animals

They range from realistic to fanciful, regal to whimsical, but all of these bells venerate members of the animal kingdom in one form or another. Perhaps it is the very diversity of our fellow creatures that leads to their frequent portrayal on bells, perhaps it is their spirit or their seemingly infinite expressions. Could they talk, it's likely most of the species represented here would be pleased with their respective representations (save, perhaps, for the pottery cat with crooked whiskers and virtually no mouth!). While two ducks and a colorful chick inhabit this virtual farmyard, bells depicting birds in general will be found in Chapter Five.

Copper bell with handle in the form of a squirrel perched on his haunches, made in New York. 5.5" high. $15-20.

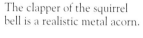

Small metal bell, probably pewter, with figural handle of pig standing on a small rectangular platform. 2.5" high. $8-10.

The clapper of the squirrel bell is a realistic metal acorn.

Brass bell with elephant handle and lotus design on base, stamped "China" inside. The bell's original owner purchased it in 1944 in San Francisco. 3.25" high. $20-25.

Two brass bells from China with figural handles representing Taoist symbols. The dove on the handle of the first bell symbolizes longevity in the Taoist religion, while the fish which serves as the handle of the other bell symbolizes abundance, fertility, strength and stamina. Left: 4" high. $10-15. Right: 3.5" high. $15-20.

A sad eyed, droopy eared dog forms the handle of this colorful ceramic bell by Jasco, part of the Critter Bell Series. Made in Taiwan. 4.5" high. $15-20.

Another Critter Bell, this one a bisque porcelain puppy sitting on a sugar bowl. Made by Jasco, 1980. 4" high. $15-20.

Glazed pottery bell from the Caribbean island of Barbados. Handle is a cat's face with green eyes and rather lopsided whiskers! 3.5" high. $10-15.

Heavy brass bell from China with figural handle of a large duck. Perched with tail up and head down, the duck's appearance is reminiscent of a fanciful dragon. 6.5" high. $35-45.

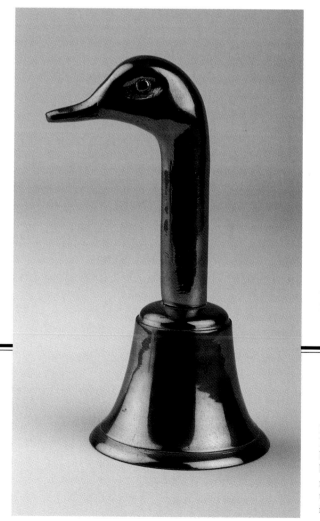

Left:
Heavy brass bell with handle in the form of a duck's long neck and head. Imprinted on side of neck: "© 1979 Custom Decor Inc." 6" high. $25-30.

Right:
Humorous version of a "dinner bell," with ceramic chick atop bell base decorated with musical notes and the announcement that "Dinner is Served." Made in Japan. 4.5" high. $8-10.

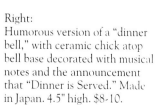

Brass bell with swimming tortoise handle, made in Taiwan. 3.5" high. $8-10.

A turtle whose shell sports a green and blue stained glass effect appears to be balancing a bit precariously on the small metal base of this whimsical bell. 4.75" high. $20-25.

Below:
Another sea horse bell, this one in clear crystal surmounted by a frosted glass handle. Two paper tags on the side of the bell read: "Golden Crown E & R France" and "Bayel Crishighin France." 7.25" high. $25-30. *Courtesy of M. "Penny" Wright.*

Brass bell with figural seahorse handle, glass eyes. 4" high. $20-25.

Silver-plated, limited edition bell from Holland with figural handle of a stalwart penguin. 3.5" high. $20-25.

Cat owners will appreciate the stylized but recognizable pose of the wide-eyed feline who serves as the handle of this heavy brass bell from Austria. A tag on the inside reads: "Handmade in Austria." 5.5" high. $15-20.

Sterling silver bell with figural handle of frog, c. 1890. The frog stands with one hand on the top of his head, the other in back. From the reverse, the figure looks like a woman with long flowing hair. 4.75" high. $50-60.

Reverse of the frog bell, showing the long flowing hair.

These four striking bronze bells are all by the same artist and feature figural animal handles. From left: cat with arched back and tail held high, 9.25" high; bird resting on twig with leaf, 5" high; flying crane with outstretched wings, 8" high; stylized dog with open mouth, 7" high. The artist's signature and a date of 1985 appear underneath the twig on the bird bell, but the artist's name is unfortunately too small to decipher. $25-30 each.

Brass bell made in Taiwan with figural horse handle. Attached to a long pole and caught in mid-stride, the nattily attired horse looks as though he just stepped off a carousel. 6.5" high. $20-25.

Old bronze bell with figural handle of man on horse carrying a mallet. Etched inside in very small letters is what looks like the word "Assyria," making it probable that this bell came from the Middle East. 3.75" high. $25-35.

Right:
Shiny brass bell with head of reindeer as handle, made in Korea. 4" high. $10-12.

Pale green celadon ceramic bell from China, figure of horse as handle. The horse stands on a circular platform which forms the top of the bell's base. 6" high. $20-25.

Expressive bronze bell by John McCombie, "The Bear Family." Cast using the Lost Wax method, this heavy, realistic bell depicts a mother bear with two small cubs clinging shyly to her legs. The clapper is a finely detailed cowboy boot. 7.25" high. $160-170.

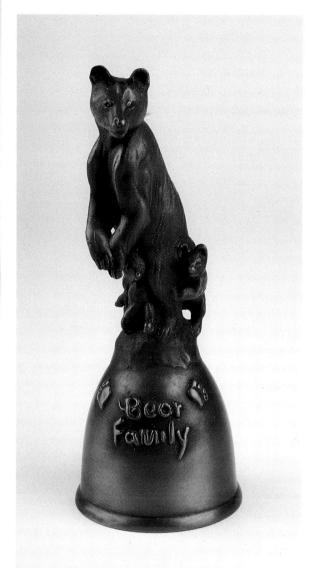

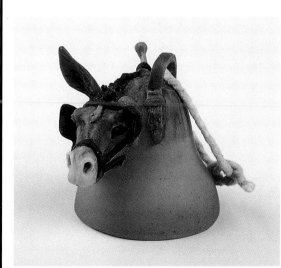

Sculptured porcelain bell by artist Lowell Davis, known for his engaging depictions of rural farm life. This one, of a mule named "Katie," is marked 1990 and signed by the artist on the back. It is part of a series that also includes a cow, goat, pig, rooster, and dogs. All are hand crafted in Scotland and distributed by Schmid. 2" high. $75-80.

Set of four colorfully decorated porcelain bells with figural handles and clappers shaped like tiny gold shells, all by artist Lynn Chase. From left: pheasant handle with matching base decoration, part of the "Winter Game Birds" series, 1992, names of all birds in the series (pheasant, woodcock, wild turkey, California quail) written inside; "Jungle Party," giraffe handle with monkey, jaguar, giant panda, and elephant decorating base, 1991; "Jaguar Jungle," jaguar handle with orange lilies decorating base, first annual bell from 1989, limited edition of 4900; "Parrots of Paradise," parrot handle and different kinds of parrots around base. All 4.5" high. $20-25 each.

The clapper for Lynn Chase's "Jaguar Jungle" bell shows the characteristic gold shell clapper.

Metal bell with rampant lion as handle. Ancient symbol of Italy's fifteenth century Medici family, the lion stands on one foot and holds an ornamental, fleur-de-lis scepter with his other three paws. 4.5" high. $25-30. *Courtesy of M. "Penny" Wright.*

Given the prolific nature of the business known as the "Bells of Sarna"—sales from the 1930s to the 1960s exceeded fifty-five million—it's hardly surprising that a string or two of these small brass bells can be found in most collections. Sajjan Singh (S.S.) Sarna was the son of an Indian dairy farmer who attended college in the United States, where he later parlayed his idea of combining small brass bells with individual story tags into a highly prosperous company. The tags provide colorful descriptions of the bells' original use in India, and although Sarna ultimately acknowledged that his stories were somewhat improvised, the images they provide of daily life in a faraway country most certainly enhance the allure of these popular bells.

The "Bells of Sarna" are amazingly diverse. They include "examples of those which Sarna first collected on his oriental tour, antique and historical bells which he bought for his private use in his home and many others he had hoped to display in a museum collection." (Schick 1981) The sets of bells strung together on multicolored strings are probably the most favored, however, and are the ones typically accompanied by the standard Sarna identification tag and small informational booklet. Whether or not one believes the tales surrounding the bells, the "Bells of Sarna" are pleasing to collect and their creator's unique place in bell history has been firmly entrenched.

String of three Sarna Cotton Beater bells on orange cord. According to Sarna's story, the job of cotton beaters in India is to thrash the cotton used for making quilts into a "fluffy down." They use a special bow with a bell attached to announce their presence and attract local customers. Graduated heights of 3.25", 3", and 2.5". $15-20. *Courtesy of M. "Penny" Wright.*

Below and below right:
String of three Sarna bells on purple cord, original Sarna booklet attached. Bells have leaf or shell like etchings and average 2.5" high. $10-15. *Courtesy of M. "Penny" Wright.*

String of five different Sarna bells on green cord, most with individual tags. The bells include a Najoomi (Fortune Teller) bell, India Elephant bell, and Chai Garam (Hot Tea) bell. $15-20. *Courtesy of M. "Penny" Wright.*

String of three Sandhu (Wandering Monk) bells on peach colored cord by Sarna. Graduated heights of 3.25", 3", and 2.75". $20-25. *Courtesy of M. "Penny" Wright.*

Interior of one of the Wandering Monk bells, showing the "Bells of Sarna" inscription.

String of five India Brass Bullock Bells, made by Sarna and inscribed "Sarna India" on reverse side. Original Sarna tag reads: "Bulls are just as holy as cows in India. They are used as beasts of burden and haul goods on a rera (a two wheeled rickety contraption). Traffic is always heavy and these bells are necessary to warn pedestrians to avoid injury by making way for the heavy and rough traffic through the muddy and dusty bazaars." All 2" high. $20-25. *Courtesy of M. "Penny" Wright.*

Elephant bell by Sarna, with original Sarna tag. The tag notes that while such bells have been worn by elephants in India "for centuries," they have found a more contemporary use in the west as dinner bells. 2.25" high. $5-10.

String of four Sarna bells with red highlights on a multicolored cord. Graduated heights of 3.75", 3", 2.5", and 2.25". $20-25. *Courtesy of M. "Penny" Wright.*

Brass bell with painted black decoration, made by Sarna. Marked around the interior of the rim: "Bells of Sarna India M31-3." 4" high. $20-25.

Size is the common denominator here: each of these bells is less than two inches high and would fit easily in your pocket or palm. Despite their diminutive proportions, however, several are intricately designed, others colorfully painted or charmingly designed. Perhaps the most interesting is the metal souvenir bell from King's Canyon National Park—easily dismissed at first glance, it takes on new meaning once the identity of the odd-looking handle is revealed.

A trio of charming miniature bells in pewter from Lunt's Silversmiths. From left: four-tiered wedding cake, decorated pepper mill with handle, flowerpot. All 2" high. $25-30 each.

Tiny ceramic bell with even tinier bird hand painted on the front. Signed by the artist, Betty Daley from Syracuse, New York. 1.75" high. $5-10.

A trio of miniature ceramic holiday bells, by Hallmark. From left: bunnies with tree, 1990; holly leaves and berries, 1991; teddy bear, 1992. All 1.25" high. $5-10 each.

Three miniature crotal bells from Japan in the form of a rooster, monkey, and tiger, all with short rope handles in contrasting colors. These bells are part of a series representing the signs of the Zodiac. All 1.5" high. $10-15 each.

Left:
Miniature brass bell from Spain, multicolor geometric decorations on base. 2.25" high. $10-15. *Courtesy of M. "Penny" Wright.*

Right:
Miniature ceramic bell with picture of two children floating on cloud and holding stars. Red tag on reverse reads: "Little Twin Stars SANRIO 1976, Made in Japan." 1.75" high. $10-15. *Courtesy of M. "Penny" Wright.*

Miniature brass bell with needle-like engraved design, by Sarna. 1" high. $3-5.

Although the "tree" perched atop this miniature souvenir bell from King's Canyon National Park is only the size of a tiny twig, it is meant to represent a majestic California sequoia. 2" high. $5-10.

One from a series depicting the "Twelve Days of Christmas," this miniature pewter bell features two bas relief turtle doves perched over a traditional Christmas poinsettia. 1.75" high. $15-20.

Musical

While some might argue that all bells play "music," this assortment of bells is musical in a more mechanical sense; each is both a bell and a traditional music box, playing its ever-present tune with the turn of a key found on the bottom.

Above two photos:
An assortment of small bells by Sarna. All approximately 2" high. $5-8 each.
Courtesy of M. "Penny" Wright.

Musical bell of white porcelain decorated with a rocking horse, assorted toys, and a pretty striped bow. When wound, the bell plays the song "Toyland." 6" high. $20-25.

Two musical bells, both made in Japan by the Westland Company. The bell on the left with poppy decoration and a peacock handle plays "Lara's Theme" from the 1965 movie *Doctor Zhivago*. The bell on the right has an iris decoration and swan handle; it plays a melody written by the Italian composer Paganini. Peacock: 5" high. Swan: 4.5" high. $20-25 each.

Lenox porcelain Christmas bell from 1992, plays "Deck the Halls" and features scene of Victorian family decorating for the holidays. The gold handle has a festive red ribbon attached. 4.5" high. $55-60.

The Reuge Collector's Musical Mother's Day bell, second issue, 1976. Made in Switzerland, this bell plays a waltz tune by Johannes Brahms. It has a tapestry effect child's picture on the front, orange velvet on the back, and gold braid trim and cord. 4" high. $25-30. *Courtesy of M. "Penny" Wright.*

Reverse of the Reuge Mother's Day bell.

Right:
Petite music box bell by Fenton, made of Ruby glass with iced snowflakes. The snowflake theme carries over to the song played by the bell: "White Christmas." 4.5" high. $25-30.

Far right:
A second petite music box bell by Fenton. This one is on Opal Satin and plays "Oh Little Town of Bethlehem." 4.5" high. $25-30.

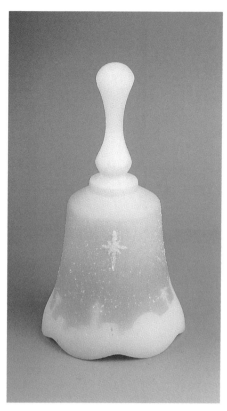

Below:
An Easter bunny and two perky chicks serve as the handles of these three Easter bells. All are made of porcelain bisque painted in traditional Easter hues of yellow, green, pink, and pale blue. Average height: 4.5". $8-10 each.

Seasonal and Holiday

Holidays bring out our natural inclination to beautify and decorate our surroundings. Bells—with their infinite variety of shapes, colors, and styles—lend themselves especially well to this universally enjoyed custom. The holiday bells featured here range from a set of playful Halloween crotals (all sharing the same toothy grin) to a finely crafted, cut glass Christmas bell by Waterford. Christmas bells, not surprisingly, are the most prevalent of holiday bells, and come from all over the world. While they can be found in many different styles, some of the most endearing are the bells and ornaments featuring jolly old Santa Claus.

Three of our four seasons are represented by these ceramic bells with appropriate figural handles. The Autumn and Spring bells were made by Roba: Autumn in 1991 and Spring in 1992. The Winter bell is inscribed inside: "Festival of Flower Fairies." Autumn: 4.25" high. $8-10; Winter: 5.25" high. $25-30; Spring: 4.5" high. $8-10.

The base of this unglazed ceramic bell is painted to resemble a decorated Easter egg. A brown bunny holding a carrot serves as the handle. 5" high. $10-15.

Two figurine bells made of bisque porcelain, representing the months of "October" and "November." October wears a black hat and carries a basket bearing a small black kitten, November holds ears of corn and carries a basket packed with seasonal fruits and vegetables. Made by Enesco, 1990. Both 5.5" high. $25-30 each.

Glazed ceramic figurine representing "Fall," skirt decorated with the falling leaves associated with this season. Tag inside reads: "Norcrest / Fine China / Japan." 3" high. $10-15.

Tiny metal crotal bells in the shape of a jack-o-lantern, a black cat, and a smiling ghost—a perfect trio for Halloween! Heights, from left: 1", 2.5", 2". $2-5 each.

Ceramic pilgrim and pumpkin bell, suitable for holiday decorating at Halloween or Thanksgiving. We either have a very short pilgrim or a very large pumpkin! 4.5" high. $20-25.

Thanksgiving bells of glazed porcelain with wooden handles, proclaiming "Let us give thanks" and "Count your blessings." Made by Autumn Treasures, 1981 and 1982. Left: 5" high. Right: 4.25" high. $8-10 each.

This 1982, porcelain bisque Christmas bell by Lefton China features a nativity scene in bas relief, hand painted in delicate pastel colors. Lettering inside identifies the bell as from the "Christopher collection." 6.75" high. $15-20. *Courtesy of M. "Penny" Wright.*

Contemporary porcelain Christmas bell with gold star as handle, made by Russ. The base is decorated with a bas relief shepherd and sheep motif on both sides. 4.75" high. $8-12.

Brass Christmas bell decorated with enameled holly leaves, made in India. 5.25" high. $15-20.

Official Bethlehem Christmas bell from 1978, "Glad Tidings."
Seals of Gregorios, Archimandrite of Bethlehem, and Bishop
Maximus of Nazareth and Galilee appear on the reverse. 6.25"
high. $25-30.

Left and above:
Porcelain Christmas bell from Sweden, shown from two
sides. This first limited edition called "The Three
Kings" was designed by Jacqueline Lynd and made by
Rorflrund, Sweden. It is trimmed in 14 kt gold and
marked inside: "Julpoesi 1980 / De Helga Kingar Tre."
5.25" high. $50-60.

Reverse of the Bethlehem
Christmas bell.

Christmas bell by Lindner, made of cobalt blue porcelain. A little drummer boy in 14 kt gold plays on the front, "Christmas 1973" appears on the back. Limited edition of 3,000, marked "Made in Western Germany." Lindner also produced Christmas bells in 1974 and 1975, both in the same cobalt blue porcelain but decorated with a little girl caroling and a sleigh ride scene, respectively. 4.5" high. $35-40.

Although not immediately apparent, this carnival glass bell features a Christmas motif: Santa with his sleigh and reindeer decorate the main body of the bell, a continuous ring of hand-holding Santas trots along the rim. 6.25" high. $35-40.

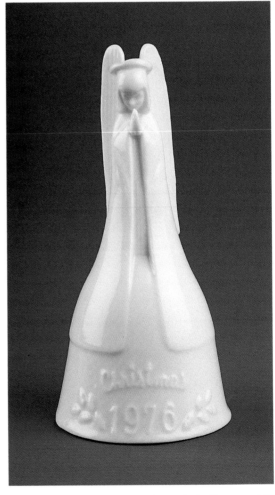

White ceramic angel figurine bell inscribed "Christmas 1976." Written inside in gold letters is: "© HM 76 594." 5.5" high. $20-25. *Courtesy of M. "Penny" Wright.*

Children prepare for Christmas on this colorful ceramic holiday bell, shown from three different sides. Marked inside: "Kaiser Germany Motiv 10." with Christmas scene, 3.5" high. $35-40.

Wedgwood bell in blue jasper with candle decoration and ring of holly leaves just below the handle. This bell is one from a series known as "Christmas Bell Ornaments of the World's Great Porcelain Houses." 2.25" high. $30-35.

Two Christmas bells in blue jasperware by Wedgwood, 1996 on the left and 1997 on the right. Although the Wedgwood company had stopped manufacturing bells of all kinds for several years, production was resumed in 1996. Both 4.25" high. $65-70 each.

Pair of Christmas bells from Avon, glazed ceramic bases with unglazed handles. The bell on the left is decorated with angels and is from 1992, the one on the right is from 1987 and features children caroling plus a snowflake handle. Average height: 5". $20-25 each.

This 1989 ceramic Christmas bell, also from Avon, features a whimsical scene of two children kissing, presents held behind their backs. Holly leaves tied with a bright red ribbon form the handle. 5.5" high. $20-25.

Ceramic figurine of little girl in Christmas garb. Paper tag inside reads: "Wolin / Japan." 3.5" high. $10-15.

Bisque porcelain "Season's Greetings" bell. A clown in holiday garb with violin forms the handle, Christmas trees and red stockings encircle the base. 4.5" high. $8-10.

Unglazed porcelain Holly Hobbie™ figurine bell, "The Music of Christmas." Limited edition, 1980. 3" high. $15-20.

Ceramic bell ornament in the shape of a happy snowman carrying a broom and—appropriately—a bell. 3" high. $5-8.

Mickey Mouse™ carrying a small Christmas tree appears on the handle of this bell distributed by the New England Collectors' Society (a division of Reed & Barton). Titled "Mickey's First Christmas," the bell is marked "NE © Disney" on the reverse. 2.75" high. $20-25.

A trio of lighthearted, contemporary holiday bells. From left: Ziggy™ bell ornament from 1982, "Joyous Holiday," 2.25" high. $5; Peanuts™ Christmas bell from 1976 featuring Snoopy and Woodstock, 5.75" high. $10-15; Santa bell ornament from 1986, by Lillian Vernon, 2" high. $5.

Pewter bell ornament in the form of a nicely detailed Santa, a sprig of holly on his cap. 2" high. $15-20.

Silver-plated second edition bell with detailed Santa figure as handle. Marked "Christmas 1997" on the front. Notice the holly leaves encircling the bottom of the otherwise plain base. 5.5" high. $10-15.

Santa can hardly be seen behind all the toys and gifts that decorate this whimsical ceramic bell! His sturdy black boots serve as clappers for the bell. 4" high. $10-15.

Figural silver bell from the Lincoln Mint, 1976. Plain silver base surmounted by figure of Santa stepping out of a chimney with sack of toys on his back. 4.5" high. $10-15. *Courtesy of M. "Penny" Wright.*

Figural silver bell from the Danbury Mint, 1978. Plain silver base surmounted by figure of waving Santa with sack of toys at his feet. 4.5" high. $10-15. *Courtesy of M. "Penny" Wright.*

This diminutive pewter bell from England depicts Santa seated atop a chimney, pulling up his sack of toys. 1.25" high. $15-20.

Pewter bell with figural handle of a realistic Santa Claus carrying a sack of toys over his shoulder. 3.5" high. $8-10.

Silver-plated bell by Gorham, with pewter handle of Santa and "Christmas 1979" inscribed on base. Is Santa hiding something behind his back? 4.25" high. $35-40.

Glass and ceramic bell ornament with colorful Santa figure as handle, Christmas stocking filled with toys as clapper. 5.75" high. $10-15.

Hand made ceramic bell ornament, rotund Santa holding a bouquet of flowers. Signed and dated (1982). 2.5" high. $15-20.

A side view reveals Santa's secret—he holds a small teddy bear destined to gladden some deserving child's Christmas.

Unglazed ceramic bell ornament by Gregory Perillo, made by Artaffects. Marked inside: "1987 Annual Bell Ornament Gregory Perillo's Sagebrush Kids / by Vague Shadows / © 1987, Ltd. / Hand crafted - Mexico." Signed by the artist on the back. 3.5" high. $25-30.

Red and white ceramic bell with stylized leaf decoration by Villroy & Boch, made in Germany. This bell is from the Bells of Christmas Series, Hamilton Collection and is marked inside: "Villroy & Boch serif 1748 Germany © 1985 HC." 2.75" high. $25-30.

Another of Gregory Perillo's Sagebrush Kids ornaments. This one is from 1986 and is also signed on the back. 3.25" high. $25-30.

Small ceramic bell with colorful Christmas scene, gold trim, and eight point star handle, also part of the Bells of Christmas Series. Around the base is written: "Nöel, Weihnachten, Christmas." The inside is marked: "Kaiser Porcelain West Germany © 1985 HC." 2.75" high. $25-30.

Perillo's signature on the back of the bell ornament.

Holly berries decorate this third small bell from the Bells of Christmas Series. It is marked inside: "Crown Staffs England © 1986 HC." 2.5" high. $20-25.

Another from the Bells of Christmas Series, this small ceramic bell with an ice skating scene is by Royal Worcester. It is marked inside: "Royal Worcester Made in England © 1985 HC." 2.5" high. $20-25.

Coalport bell, also from the Bells of Christmas Series, poinsettia decoration on front. Marked inside: "Coalport Made in England © est. 1750 © 1986 HC." 2.25" high. $20-25.

1992 Christmas bell of clear cut glass, made by Waterford. Holly leaves surround the figure of a lithe young dancer on the front. 5.35" high. $75-05.

141

Mark inside the Bing &
Grondahl bell.

Above and above right:
Blue porcelain Christmas bell from Denmark
by Bing & Grondahl, with doves of peace
design and star finial. This bell is part of a
Christmas bell series from Bing & Grondahl
modeled after the company's plates. It is
marked "Jule After 1897" on the reverse and
"Bing & Grondahl Christmas Bells" inside.
3.5" high. $25-30.

Blue porcelain, limited edition *Arsklokke*, or Year Bell from Denmark, by
Bing & Grondahl. One of a series of bells all honoring cathedrals from
around the world, the base decoration of this 1981 bell illustrates Upsala
Cathedral in Sweden. Around the rim is inscribed: "Ring out the Old, Ring
in the New, Ring out wild Bells," while around the shoulder is inscribed:
"Uppsala Domkyrka Sverige, 1981." The clapper is a wooden ball. A most
elegant way to celebrate New Year's Eve! 5.25" high. $45-60.

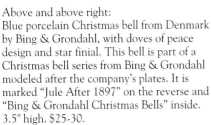

Another bell from the Bing & Grondahl
Christmas series. This one has a similar base
but different style handle. It reads "Jule After
1983" on reverse. 3" high. $25-30.

Bell jewelry is hardly new, despite its status as a favored adornment for many contemporary bell lovers. Throughout the years and throughout the world, bells have bedecked toes, ankles, wrists, necks, and earlobes. Although the wearing of bells occurs most often for ornamental reasons, it sometimes serves the dual purpose of alerting others to the wearer's presence. In Arab countries, for instance, servants reportedly know when their mistress is approaching by the sound of the jingling bells suspended from the latter's clothing. (Mayer 1992, S-11)

In our society, however, it's likely that those who fancy bell jewelry do so simply because it is pretty and festive, a visual affirmation of one's ongoing affinity for bells. The pleasure derived from wearing bell jewelry has been quite aptly described by artist Gordon Barnett. In a catalog illustrating his hand crafted bell jewelry, Barnett advises prospective purchasers that all the bells "can be worn in some manner, although each begs to be put to use in other creative fashion. . . to become as frivolous or ceremonial as you wish. They just want to have fun, to be imaginatively meaningful—to be loved."

Angel stickpin decorated with tiny, functional bell, also by Terry Mayer. 2.5" long. $100-200.

Bell pendant, with clapper comprised of two winged cherubs climbing up a rope, by Terry Mayer of New York City. $100-200.

Bell shaped mother-of-pearl pendant on chain. 1" high. $15-20.

Sterling silver and enamel pendant in
bell shape with Siamese dancer on front.
$25-35.

Pair of sterling silver, open work
bell charms with functional clappers
on chain. $15-20 each.

Bell shaped cloisonné pendant on
chain, functional clapper inside. 1.5"
high. $20-30.

Bow and bell brooch with three
tiny but functional bells. 2" long.
$20-30.

Two festive bell brooches, one with basketwork design, the other accented with red. $10-15 each.

Bell shaped brooch with faceted blue clapper and bow at the top. $10-15.

Silver pendant adorned with three small bell charms. The Canterbury bell charm on the left and the hand bell charm in the center were both purchased at the Taylor Bell Factory in Loughborough, England, one of two bell founders remaining in England. The third charm is unidentified, but made of silver. $75-85.

Steeple shaped bell pendant in sterling silver by Gordon Barnett, made in the Lost Wax method. 1.5" high. $65-75.

Charm bracelet with eight miniature bells, c. 1950s. Four of the bells are open mouth type with clappers, the other four are crotals. Value undetermined. *Courtesy of M. "Penny" Wright.*

Sterling silver, cowbell shaped pendant decorated with poinsettia-like flower. $25-35.

Lighthouse shaped bell pendant with functional clapper. $50-60.

Bronze pendant in the shape of a Canterbury bell flower, complete with stamen shaped clapper. 1.5" high. $70-80.

Sterling silver bell pendant by Gordon Barnett, shaped like a crocus flower with realistic, stamen shaped clapper. $75-90.

Trio of miniature cloisonné bells forming a colorful pendant. Two of the bells have turquoise enamel interiors, the third is royal blue inside. Different color beads serve as the clappers. $15-20 each.

Elaborate necklace festooned with tiny crotal bells, most attached to short cylindrical rods separated by seven pale green stones. $55-65.

Left and far left:
Pendant made from one of the "Lucky Bells of San Michele." These bells were brought home by World War II servicemen from the Isle of Capri off the coast of Italy. $20-40, depending on size.

Right and far right:
Ornate necklace with a multitude of tiny crotal bells hanging from the large circular pendant in the center. $25-35.

Chapter Five

Melodious Motifs

Although their perennial diversity is one of bells' most fascinating aspects, there are nonetheless many common themes allowing them to be grouped together for display or study. Some collectors enjoy focusing on one particular motif or another, a practice that enhances their enjoyment of "the hunt" for that special bell meriting a place of honor on the shelf or mantel. Four such motifs—birds, flowers, children, and novelty clappers—are illustrated here. They represent, of course, only a sampling of the many choices available; the list of other potential themes is limited solely by the collector's imagination and can be generated by a special interest or hobby, a profession, heritage—perhaps even the color scheme of the room in which the bells are displayed!

Among the bells presented here, the respective motif may show up as either base decoration or handle design and may be a realistic, stylized, or even humorous portrayal of the selected subject matter. Sometimes a specific type of bird or flower can be identified, for instance, at other times the decoration may appear more of a composite or even a fanciful depiction. The children's bells include those using figures or drawings of actual children as well as those featuring related items, such as toys, teddy bears, or characters from children's literature. And the bells with novelty clappers need no real introduction—their creativity and charm speak easily for themselves.

Birds

A bird with outstretched wings standing on a five petaled flower serves as the handle of this small brass bell. 3.75" high. $12-15. *Courtesy of M. "Penny" Wright.*

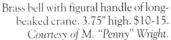

Brass bell with figural handle of long-beaked crane. 3.75" high. $10-15. *Courtesy of M. "Penny" Wright.*

Far left:
Heavy old brass bell with solemn looking eagle as handle. The handle is flat, one sided style with "China" etched on the eagle's back. 5" high. $15-20.

Left:
Metal bell from Germany with hand carved and hand painted wooden parrot handle, c. early twentieth century. The parrot is marked on the back: "Germany." 4.5" high. $15-20.

Below:
Set of pewter wind chimes comprised of six fluttering hummingbirds attached by black string to a central bell shaped cup. 12" long from top of cup to birds. $15-20.

Far left:
Metal bell with figural handle of nesting bird, from Persia. The bird sits on a bright blue bead with a "nest" made of yarn beneath. Tooled designs decorate the base. 6.875" high. $30-50.

Left:
Unglazed pottery bell from Mexico with brightly colored bird as handle, hand painted sunburst pattern around base. 6.75" high. $15-18.

Right:
This heavy brass bell with somber bronze owl handle apparently honors both a person and an event. In faint script on the front is inscribed the name "Gene," and on the back is written: "1948, 1979 N.C.C." 3.75" high. $12-15.

Far right:
Heavy brass bell with feathered owl handle, purchased in Ireland but most likely made in England. 4.5" high. $12-15.

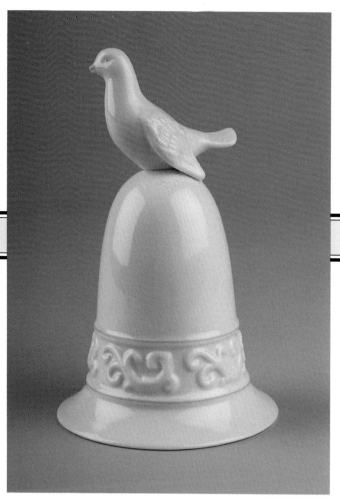

Two doves form the handle of this white ceramic bell, one with wings spread wide, the other with wings folded. Tag inside reads: "Lefton's Reg US pat. Off. exclusives JAPAN." 4" high. $8-10. *Courtesy of M. "Penny" Wright.*

Doves are a well known symbol for peace, and the pure white dove atop this glazed ceramic bell by Avon lends a serene and graceful appearance to the bell. Slender orange trim around the scrollwork border adds a touch of color. The inside reads: "Avon / The Tapestry Collection / Exclusively Handcrafted for Avon 1981." 5" high. $15-20. *Courtesy of M. "Penny" Wright.*

Left:
French faïence bell, hand painted figural handle of bird eating a cherry. Marked inside: "Mottahedel Hand painted Creations Ltd. France." 4.5" high. $45-60.

Right:
Glazed ceramic bell with decorated loop handle, made in Italy. The base illustrates a yellow bird with upturned beak, fluffy blue wings, and a red tail. 3.25" high. $10-15.

Hand painted bone china bell by Herend of Hungary, with famous Rothschild bird forming the decoration on the base. The string of jewels hanging from a branch between the two birds represents the treasures of the great Rothschild banking family. Marked "Herend Hungary / hand painted" inside. 3.5" high. $40-60.

Glazed porcelain bell with gold loop handle, decorated with flowers, leaves, and a redheaded bird. Tag inside reads: "Shefford Japan." 4" high. $15-20.

Glazed porcelain bell with graceful loop handle by the Hutschenreuther Company of Germany. The base is decorated with pink flowers, gray leaves, and a small brown bird. The word *Januar* is German for "January." 4.25" high. $25-35.

Bone china bell from England by Minton. The central motif is an aqua and rose peacock with wings folded. The bell is also decorated with pastel flowers and gold trim. 5.25" high. $35-40.

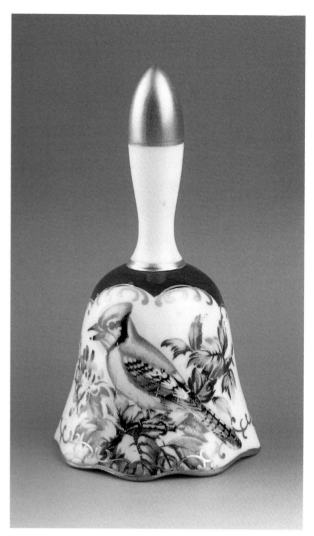

Bone china bell with gold tipped handle. The subtly scalloped base is decorated with royal blue, exotic looking flowers and a bright eyed bluejay. 5.25" high. $15-20.

Cowbell shaped porcelain bell, pale yellow with hand painted decoration of two petite birds sitting on a tree branch. The reverse side is decorated with a house in a pastoral setting. Stamped inside: "Made in Germany." 3.5" high. $60-75.

Far left:
Glass bell by Fenton, a limited edition from the 1989 Connoisseur Collection. Bluebirds are hand painted on the Rosalene Satin bell body, which subtly changes in color from warm pink to pale white. Prior to 1977, Fenton bells in the Rosalene color had a sharper delineation between the pink and white colors. 7" high. $25-30.

Left:
Lead crystal bell with faceted crystal clapper, topped by a realistically sculpted porcelain goldfinch by artist Wilhelm Buehler. This bell was sold by the Franklin Mint in 1986 and advertised as the "world's first porcelain crystal bell." Subsequent birds in the series include a blue tit, black-capped chickadee, and nuthatch. 5.5" high. $75-80.

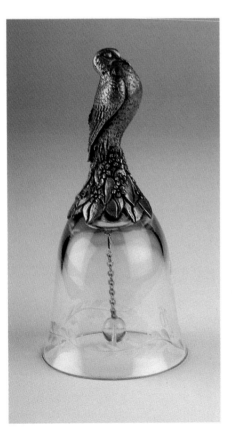

Far left:
The handle of this glass bell from the Franklin Mint is comprised of a pewter dove with wings folded back and chest puffed out. He stands amid a cluster of flowers and leaves contoured to the shape of the bell's top. The glass base is etched with a leaf and vine decoration and the date "1976." 5.75" high. $60-70. *Courtesy of M. "Penny" Wright.*

Left:
A graceful swan is incorporated into the handle of this crystal bell from the Danbury Mint. It is accented with gold trim around the base and a gold chain for the clapper. Tag on the back reads: "Sasaki Crystal / Hand Crafted Japan" with 1980 printed underneath. 7.75" high. $50-55. *Courtesy of M. "Penny" Wright.*

154

This unusual bell in the shape an Easter lily comes from the island of Bermuda and was supposedly made of cast iron from a ship that sunk off Bermuda's coast. The iron was mixed with copper and bronze to give the bell its coloring. English registry mark found inside. 4.75" high. $65-75.

Bohemian crystal bell in cobalt blue with a gold overlay and enameled pansies. 4" high. $35-40.

Tall brass bell from India, enamel overlay on base of purple and rose flowers accented by green leaves. Tag on inside reads: "Made In India." 7.75" high. $10-15.

Far left:
Glass bell by Fenton, "Blue Royale," translucent cobalt blue accented with hand painted orange roses. 6.75" high. $20-25.

Left:
Glass bell by Fenton, "White Roses on Ruby" in Medallion shape. This bell was first sold in 1979. 7" high. $25-30.

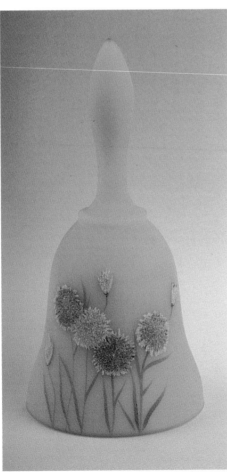

Right:
Glass bell by Fenton, "Frosted Asters on Blue Satin." This pastel colored bell was sold by Fenton from 1973-1984. 6.75" high. $20-25.

Far right:
Glass bell by Fenton, "Decorated Violets" on Spanish Lace Petticoat glass. Produced from 1974-1984, this bell is also known as "Violets in the Snow." The sparkling glass treatment on the bell's ruffled rim is called Silver Crest. 6.5" high. $20-25.

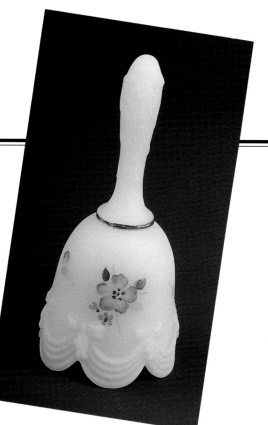

Petite glass bell by Fenton,
"Kristen Floral," Bow and
Drape shape on Opal
Satin. 4.5" high. $18-20.

Delicate glass bell with hand painted
white rose design on front, crystal
handle. 5.5" high. $25-30.

Another bell by
Fenton in the Aurora
shape, this one in
black with hand
painted pink and
yellow flowers by
D. Robinson. It is
known as
"Victorian
Bouquet." 7"
high. $25-30.

Glass bell by Fenton in Aurora shape,
blue with hand painted yellow and white
flower decoration on front. 7" high. $20-25.

Glass bell with hand painted iris decoration and a long slender handle. 4.5" high. $15-20. *Courtesy of M. "Penny" Wright.*

Tall glass bell with swirl handle, hand painted flower and bow design over white base. Tag on back reads: "Norleans / Hand Made in Italy." 7.75" high. $15-20.

Hand painted porcelain bell from Italy in the shape of a country flower basket filled with pink and yellow roses, matching yellow bow atop the handle. Tag on the bell reads: "Nuova Capodimonte." 6.75" high. $40-50.

Three ceramic bells from the Danbury Mint's American Flowers Bell Collection. From left: "Corn Poppy," "Jonquil," and "Sweet Peas." All 4.5" high. $20-25 each.

Hand painted bisque china bell from the Franklin Mint with large, realistic rose as handle, green leaves and stems draped over the white bell base. This limited edition bell by floral artist Jeanne Holgate is called "The Sonia Rose" and was created in 1983. 4.75" high. $65-75.

A similar flower basket motif is found on this second bell, also marked "Nuova Capodimonte." It has a large pink rose in the basket and a tiny blue flower atop the handle. 6" high. $35-40.

Bone china bell by Royal Albert with delicate spray of pink, lavender, green, and yellow flowers topped by a heart shaped finial. Marked inside: "Royal Albert® / Bone China England / "Fragrant Flowers" © 1990 Royal Albert Ltd." 3" high. $15-20.

Inside of the Royal Albert bell, showing the clapper and inscription.

Porcelain bell of Irish Belleek, with pale pink flowers, green shamrocks, and a gracefully scalloped rim. 4.25" high. $65-75.

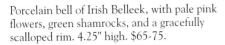

Porcelain bell with pink blossoms on both sides, gold trim around the rim and base of the handle. 4.75" high. $15-20. *Courtesy of M. "Penny" Wright.*

Porcelain dinner bell with 14 kt gold handle and trim, wooden clapper. The flowers and leaves on the front are hand painted in subtle shades of green. 4" high. $50-60.

Old porcelain bell from Vienna, Austria with hand painted pink roses, trimmed with 24 kt gold. 4.25" high. $25-30.

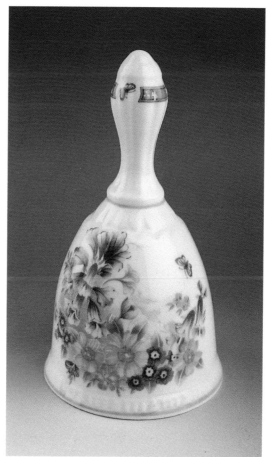

"Arabella" is the pattern name for this bone china bell by Royal Doulton with pastel colored flowers and butterflies decorating the base. It was purchased in Stoke-on-Trent, England and is marked inside: "© 1990 Royal Doulton." 4.75" high. $25-30.

Bone china bell by Royal Albert, made for the Danbury Mint's series called "Bells of the World's Great Porcelain Houses." It has a floral bouquet decoration both front and reverse and a curved handle resembling the stem of a flower. 4" high. $25-30.

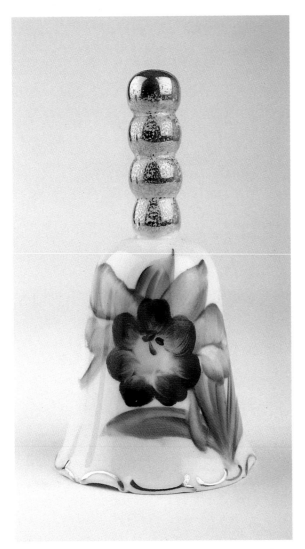

Glazed porcelain bell by Meissen, with hand painted crocus decoration on front and Meissen blue crossed swords trademark inside. 4.75" high. $50-60.

Hand painted porcelain bell with flamboyant pink and green floral motif, four stacked gold balls as handle. Unmarked, believed to be old. 5.75" high. $20-25.

A rose garden of bells, all in fine bone china from England. From left: "Hathaway Rose," by Wedgwood. 4.5" high. $30-35; yellow rose with heart finial, by Golden Crown. 4" high. $25-30; "Old Country Roses," by Royal Albert. 4.25" high. $25-30.

A second trio of bells, all with delicate violets design. From left: ceramic bell from Japan with subtly scalloped edge and gold trim. 5" high. $10-15; contemporary bell from Japan made in the style of R.S. Prussia. 3.5" high. $25-30; "Victorian Violets," by Hammersley. 5.5" high. $20-25.

Glazed porcelain bell trimmed with 24 kt gold, by Herend of Hungary. A hot pink, hand painted floral design is the primary feature of this contemporary bell. 4.25" high. $75-80.

Far left:
Bone china bell from England, by Coalport, made for the Danbury Mint's series called "Bells of the World's Great Porcelain Houses." Applied floral decoration on front, gold trim around base of handle and rim. 4" high. $25-30.

Left:
Two red roses on the front, a slightly scalloped rim, and touches of gold trim nicely accent this glazed white ceramic bell. 4.25" high. $20-25. *Courtesy of M. "Penny" Wright.*

Bottom two photos:
Bone china bell from England by Royal Crown Derby, "Derby Posies," shown front and reverse. The columnar shaped handle with gold trim complements the octagonal base. 5.25" high. $60-65.

Bone china bell from England by Royal Worcester. All white bell with floral bouquet on front, butterfly handle, and slender gold trim, 3.5" high. $20-25.

Right: Glazed porcelain bell by R.S. Prussia, white with pink and yellow floral design, scalloped edges, and a hint of gold trim. 4" high. $45-50.

Porcelain bell from England decorated with delicate blue flowers around both the rim and handle. 5.25" high. $20-25. *Courtesy of M. "Penny" Wright.*

Glazed porcelain bell by Hutschenreuther of Germany with floral decoration on base and an interesting, gold trimmed handle. 4.5" high. $30-40.

A pair of blue and white Calla lilies grace the front of this ceramic bell with open loop handle. Made in Japan for the Russ Co., Oakland, New Jersey. 4.5" high. $18-20.

Maroon and white souvenir bell from New England, matching floral design on base and handle. Marked inside: "Charlotte Royal Crownford Ironstone England." 4.25" high. $15-20.

Glazed porcelain bell from France, pale blue with gold trimmed loop handle. A large pink zinnia surrounded by smaller flowers is centered on the front in a gold trimmed white panel. 4" high. $30-35.

Brass figurine of little girl with outstretched arms, wearing a long dress and bonnet. Small label inside reads: "Made in India." 4.5" high. $15-30.

Figural bell of little girl with long wavy hair, one hand upstretched. She wears a ruffled dress tied in the back with a bow. The base of this brass bell is decorated with interesting designs and is marked "Syria" in capital letters at the top, just below the handle. 4.5" high. $45-50.

Contemporary brass figurine bell of young girl with a wide skirt, carrying a basket or purse in her left hand. Very heavy with no marks. 3.5" high. $15-20.

Left and far left:
Two figural bells from the Danbury Mint showing children preparing for bed. The young boy with candle and teddy bear is from 1976, the girl with candle and doll from 1977. Both 4.5" high. $25-30 each. *Courtesy of M. "Penny" Wright.*

Below:
These four endearing bells from Italy feature silver bases made by Reed & Barton and colorful Anri figurines as their handles. All 4.75" high. $20-25 each.

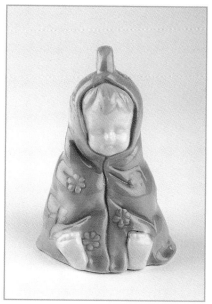

Glazed blue and white ceramic figurine of baby wrapped snugly in blanket with just his head and feet showing. 4.5" high. $10-15. *Courtesy of M. "Penny" Wright.*

Pair of annual Hummel bells by Goebel of West Germany, both with bas relief images of Hummel figures decorating the front. The bell on the left is a Second Edition from 1979 called "Farewell," the one on the right is a Third Edition from 1980 called "Thoughtful." Both 6" high. $60-70 each.

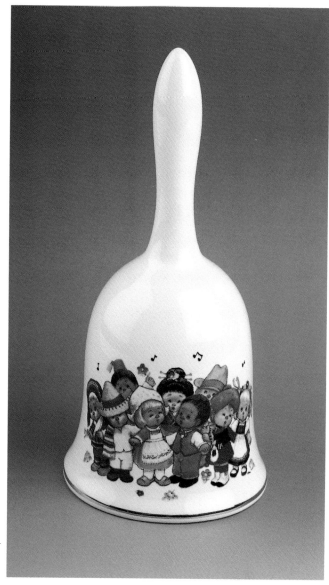

Right:
Glazed porcelain bell with charming group of singing children, entitled "Friends Around the World." Marked on inside: "© Morehead, Inc." 5.75" high. $18-20.

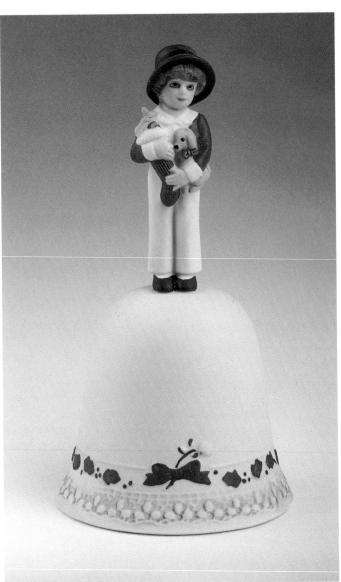

Above and above right:
These two bisque porcelain bells with handles in the form of endearing children are from a set of twelve by Texas artist Jan Hagara. "Carol" and "Chris" are both limited editions from 1986-87 with clappers in the form of a small porcelain teddy bear. Marked inside: "Distributed by Royal Orleans B & J Art Design." $95-100 each.

The teddy bear clapper from the Jan Hagara bells.

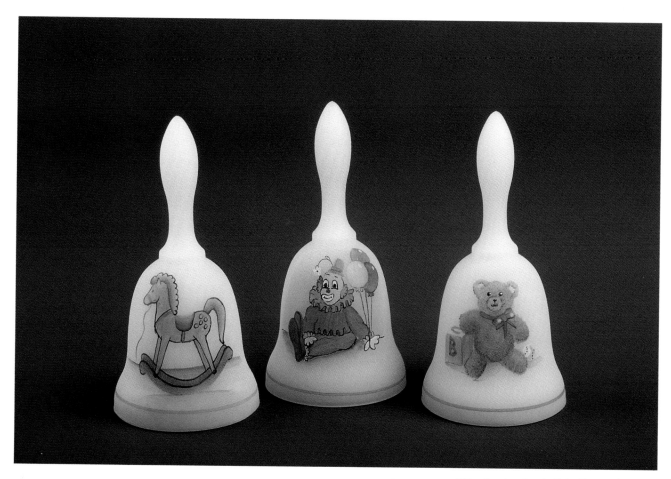

Trio of petite glass bells by Fenton from the Childhood Treasures series, all hand painted on Custard Satin glass. From left: Hobby Horse, 1984; Clown with Balloons, 1985; Teddy Bear, 1983. All 4.5" high. $18-20 each.

Scenes and words from Beatrix Potter's *The Adventures of Peter Rabbit* highlight this ceramic children's bell from England by Wedgwood. 4.5" high. $35-40.

Peter Rabbit himself sprints along the side of the Wedgwood children's bell.

Above left:
This Precious Moments™ bell and the others on this page are from a limited edition series distributed by Enesco. They are all in unglazed bisque porcelain with individual names. These and other Precious Moments™ images also appear on items such as plaques, mugs, holiday ornaments, picture frames, music boxes, and more. This graduate bell from 1981 is called "The Lord Bless You and Keep You." Note the tiny bird perched on the end of the little boy's diploma. 5.5" high. $35-40.

Above:
A boy and his pup share tears in "God Understands," 1980. 5.5" high. $35-40.

Left: "The Purr-fect Grandma," 1981.
Right: "Mother Sew Dear," 1981. Both 5.5" high. $35-40 each.

Glazed blue and white ceramic figurines of Dutch boy and girl. It is unusual to find figurines depicting male characters; note how the boy's billowy trousers have been stylized to accommodate the shape of the bell while still conveying the impression of pants. 5.25" high. $18-20 each. *Courtesy of M. "Penny" Wright.*

Ruby glass bell by Fenton, "Butterfly Delight," depicting a young girl with three fluttering butterflies. The white enamel silhouette design is known as "Mary Gregory," after the late nineteenth century American artist who first created it at the Boston & Sandwich Glass Works. 6.75" high. $25-30.

Glazed ceramic figurine bell of a little girl with her hair in pigtails. The bright gold bows in her hair match her shoes and her skirt is trimmed at the bottom with an applied pink ruffle and flowers. Made in Japan. 3" high. $15-20.

Series of bisque porcelain figurines called Merri-Bells, made in Taiwan. Tag inside reads: "© Jasco, 1978." All 4.5" high. $5-8 each. *Courtesy of M. "Penny" Wright.*

These two unglazed ceramic bells with wooden handles are from Sandstone Creations in Phoenix, Arizona. The paintings of Indian children on the front are by well-known artist Ted De Grazia. Both 5" high. $25-30 each.

Novelty Clappers

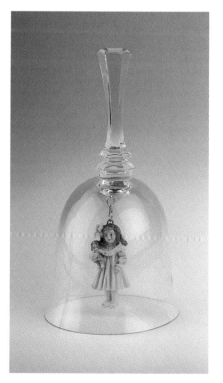

Right:
Suspended by its tiny chain, the hand crafted cherub forming the clapper of this crystal bell seems to be dancing in air. Bell made in the United States, clapper made by Fontanini of Italy. 7.25" high. $25-30.

Crystal bell with Parian porcelain girl holding doll as clapper, crafted at Intagloi Designs Ltd. 6" high. $25-30.

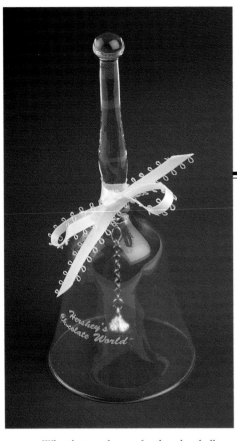

What better clapper for this glass bell from "Hershey's Chocolate World" than a tiny replica of the much-loved Hershey's kiss? A pretty white bow accents the clear glass bell nicely. 6.25" high. $15-20.

This glass souvenir bell from Vermont features a tiny but realistic covered bridge as the clapper. 4.75" high. $15-20.

This simple glass bell is adorned with a red velvet bow. A Victorian style Santa hangs merrily as the clapper. 4.75" high. $20-25.

Santa holding a miniature Christmas tree forms the clapper of this lead crystal bell with gracefully tapering handle. 6.25" high. $25-30.

Chapter Six

Bells and Beyond:
Multi-purpose and Kindred Spirits

This chapter highlights an assortment of useful or decorative items that either incorporate bells or are shaped like bells. For bell collectors, such items offer an opportunity to expand their collections into the "bell-related" realm and to learn even more about the fascinating ways in which bells can be used. For some, the acquisition of a multi-purpose or bell shaped item may even lead to the development of a second or third collection, one separate from but certainly related to bells!

Souvenir spoons, for example, are a good illustration of how one collection can easily dovetail into another. Dating originally from the late nineteenth century, souvenir spoons come in a variety of different metals and feature all manner of engaging designs and embellishments. They are collected "for the same reasons that bell collectors acquire bells: for their beauty, for their rarity, to show where [the collector] has been . . . to show off to their fellow collectors." (Brophy and Simms 1997, S2) While one might opt to collect souvenir spoons of many kinds, a collection of souvenir spoons with bell motifs would nicely

complement a bell collection. The three spoons shown here on page 179 illustrate two of the most typical bells found on these interesting collectibles: Philadelphia's Liberty Bell and California's mission bells. While all three feature the bell motif on the front of their handles, spoons are also found with bells on the reverse of the handle or in the bowl of the spoon.

Three bells with can opener handles, all made by Sarna. The tallest one on the left sports drink related phrases like "Cheerio;" the other two are decorated with characteristic Indian leaf designs. From left: 7.25" high, $15-20; 3" high, $5-8; 4.5" high, $10-15.

Bell and cup combination by Sarna, brass with star shaped painted decorations. 4" high. $10-15. *Courtesy of M. "Penny" Wright.*

Old and heavy brass tap bell with dual clappers, one on each side, originally used on the desk of an English inn. The cylindrical receptacle on top is used to hold matches, which can be lit by striking them on the rough outer surface of the cylinder. The lower portion of the bell serves as an ashtray. 5" high. $90-100.

Telephone dialer for rotary phone with miniature bell on one end, inscribed "Sarna Brass / India 5473" on one side and "Vizcaya" on the other. 4" long. $5-10. *Courtesy of M. "Penny" Wright.*

Combination bell and thermometer, with handle in the form of London's famous "Big Ben." 7" high. $15-20.

Trio of souvenir spoons with bell motifs on their handles. From top to bottom, the spoons are from the San Gabriel Mission in Los Angeles (with tiny bells hanging in the mission windows); Philadelphia, Pennsylvania (with the Liberty Bell depicted); and the state of California (with El Camino Real bell showing date of 1769). The California spoon has what is known as a pointed, cut-out handle. 5" average length. $20-25 each.

179

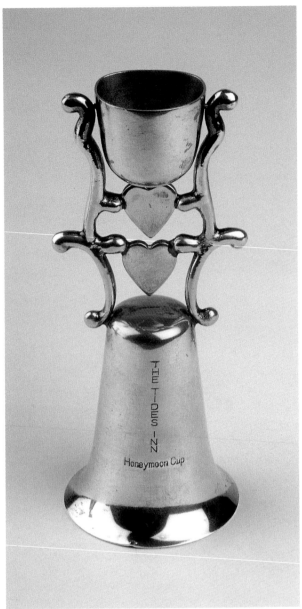

Silver bell with two small hearts and a wine cup on top, a more modern version of the much fabled wedding cup. Engraved on the front: "The Tides Inn Honeymoon Cup." 5.75" high. $25-30.

Silver embossed, highly embellished wedding or marriage cup with gold inlay. The wedding cup consists of two bell shaped cups—a larger one formed by the skirt of a woman and a smaller one held aloft by the same woman's arms. At weddings, a bride and groom both drank from the cups at the same time, the groom from the larger and the bride from the smaller. Legend holds that if both succeeded in emptying the cup without any spills the marriage would similarly succeed. 7" high. $75-100.

Some of the most highly prized multi-purpose bells are the Victorian silver-plated serving pieces that date primarily from the second half of the nineteenth century. During this era, it was considered quite fashionable to appoint the dining room table with elegant castor sets, napkin and spoon holders, toast racks, and the euphemistically named "waste bowls" (which were used to collect the liquid remnants of the meal). Often, small bells were incorporated into the top or bottom of these fancy serving pieces; they could then be dually employed for summoning the household servants.

Victorian, silver-plated waste bowl with tap bell on top. The four-footed bowl is elegantly decorated with flowers and leaves, while the heads of bearded men resembling Vikings appear at the junction of each of the legs with the bowl. 8" high. $150-175.

Elegant silver castor set with five original glass bottles and stoppers, c. late nineteenth century. The ornately decorated handle contains a holder for the separate bell which rests in the center. Lazy Susan type frame covered with engravings, stamped "Meriden B. Company." 15" high. $175-195.

Another castor set, this one with six bottles surrounding a centrally placed tap bell, c. 1860-80. Marked on the bottom "Wilcox Silverplate Co." 15" high. $175-195.

Silver-plated spoon holder with slots for twelve spoons and tap bell at the very top. Bottom of base marked: "Simpson Hall Miller & Company Wallingford Pat." 11.75" high. $150-175.

Silver sugar bowl with tap bell on cover, shown with matching creamer. The finial of the tap bell is a silver peacock with finely detailed feathering. Both pieces marked "Meriden B. Company." Sugar: 8" high. Creamer: 5.5" high. $225-250.

Left and above:
Dainty brass saccharin holder trimmed with imitation pearls. The top is a small bell, presumably used to summon lunch or dinner guests to the table. 3.25" high. $10-15.

The Victorian era is also the source of another bell-related collectible: the practical as well as pretty "smoke bells." Bells in shape only, these interesting items were originally used above early hanging lamps to protect the walls and ceilings from the oily smoke and soot given off by the lamps. They were generally attached to the lamp via a metal frame, hung high enough for air to reach the lamp's flame but low enough to keep smoke from soiling the ceiling. Smoke bells vary widely in their appearance: they can be made of glass, metal, or porcelain and range in shape from flat and saucer-like to tall and conical. Their rims may be fluted, flared, softly scalloped, or left plain. Glass smoke bells (the most common kind) were sometimes etched, frosted, cut, or cased; occasionally their designs were coordinated with the design on the lamp.

As lamp styles and lighting technology changed, smoke bells fell into disuse. Rather than discard them, however, some owners—no doubt inspired by the beauty of the bell-like shape—found new applications for their treasured pieces. Lois Springer recounts several "charming vignettes" related to the destiny of Victorian smoke bells:

> One English-woman took hers to cover her wineglasses, thus creating bell-like covers. Another story comes from Paris, where in a French movie a conical smoke bell appeared as a slow but well-timed and efficient way of snuffing out a candle. The heroine was seen to place a smoke bell over her bedside candle flame, giving herself just enough time to get into bed before the light extinguished itself from a gradual lack of air. (Springer 1972, 226-227)

Unusual, hand painted glass jar in bell shape, small loop handle at the top. The jar is hinged about two-thirds of the way up, allowing the top to remain attached when the jar is opened. 4.75" high. $40-50.

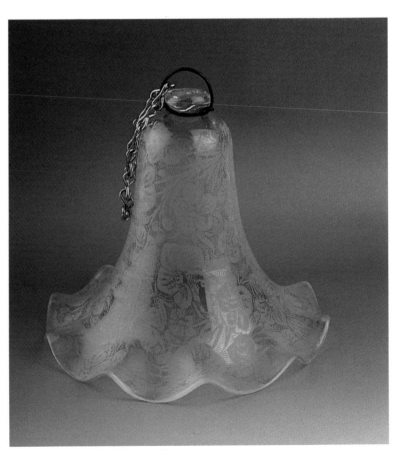

Glass smoke bell with scalloped edges and all over etched flower design, 6" high, 6" dia. $60-65.

Tall white smoke bell, made of milk glass with scalloped edges and ruby rim. Smoke bells were used above hanging lamps in the late nineteenth century to protect the walls and ceiling from soot. 9" high. $85-90.

Another milk glass smoke bell, this one all white with a flatter profile and fluted edges, c. 1900. 3.5" high, 6.5" dia. $40-45.

187

The American Bell Association

Those new to bells and those with an existing collection who want to expand their knowledge and understanding of bells will both find a valuable resource in the American Bell Association (ABA). This non-profit organization was founded in 1940 and has over forty-five regional, state, and international chapters that meet on a regular basis. The group holds an annual convention every June and publishes *The Bell Tower*, a bimonthly magazine featuring articles on all kinds of bells, collecting tips, pricing information, and chapter news. Some ABA members are particularly knowledgeable about certain kinds of bells—large bells, glass bells, or sleigh bells, for example—and are willing to share their experience and expertise through articles submitted to *The Bell Tower* or individual discussion with other members. The convention, held in a different location each year, affords bell collectors the opportunity to attend informational programs and develop lasting friendships with other bell fanciers. A highlight of each convention is the annual bell auction.

For further information, write to:

ABA
P.O. Box 19443
Indianapolis, IN 46219

The American Bell Association Organizational Bell, first issued in 1983 and sponsored by the Past Presidents of the Association. Available for members of the organization to purchase, this pewter bell was designed by Robert Lamb of Gatehouse Studios, Lincoln, Rhode Island, and produced by Hampshire Pewter, Wolfeboro, New Hampshire.
It is engraved with the official ABA logo, including 1940 (the year ABA was established) and the name of the calligrapher, R. F. Dabol. Each hand cast bell is signed and numbered. 6" high. $45-50.

Bibliography

Anthony, Dorothy Malone. *The World of Bells*. Des Moines, Iowa: Wallace-Homestead Book Co., 1971.

——. *World of Bells No. 2*. Des Moines, Iowa: Wallace-Homestead Book Co., 1974.

——. *World of Bells No. 5*. no date.

——. *Bell Tidings*. no date.

——. *The Lure of Bells*. 1989.

——. *World of Collectible Bells*. no date.

——. *More Bell Lore*. 1993.

——. *Bells Now and Long Ago*. 1995.

——. *Legendary Bells*. 1997.

"A Bell Bio—With Bells On." *The Bell Tower* 50, no. 1 (January-February 1992): 10.

"Additional Bells in Color." Compiled by The Bell Tower Standing Committee. *The Bell Tower Supplement* 55, no. 2 (March-April 1997) S9-S12.

Baker, Donna. *Collectible Bells: Treasures of Sight and Sound*. Atglen, Pennsylvania: Schiffer Publishing, Ltd., 1998.

Brenner, Robert. *Depression Glass for Collectors*. Atglen, Pennsylvania: Schiffer Publishing, Ltd., 1998.

Brophy, Elaine, and Peggy Simms. "Souvenir Spoons with Bells On." *The Bell Tower Supplement* 55, no. 6 (November-December 1997): S2-S14.

Chellis, Elizabeth. "Wedgwood Bells." *The Bell Tower* 56, no. 3 (May-June 1998): 16-17.

Childress, Phillip. "The Legend of the Lucky Bells of San Michele From the Isle of Capri." *The Bell Tower Supplement* 54, no. 1 (January-February 1996): S10-S20.

Florence, Gene. *Elegant Glassware of the Depression Era*. Paducah, Kentucky: Collector Books, 1993.

Glassco, Marjorie. *An Introduction to Bell Collecting*. The American Bell Association International, Inc., 1992.

Grant, Joyce. *NY World's Fair Collectibles, 1964-1965*. Atglen, Pennsylvania: Schiffer Publishing, Ltd., 1999.

Hamlin, Gene. "Basics for Beginners." *The Bell Tower* 52, no. 1 (January-February 1995): 24-26.

Heacock, William. *Fenton Glass: The Third Twenty-Five Years*. Marietta, Ohio: The Glass Press, Inc., 1989 and 1994.

Hites, Beatrice. "China vs. Porcelain." *The Bell Tower Supplement* 52, no. 6 (November-December 1994): S2-S16.

Hogan, E. P. "19th Century Call Bells in Silverplate." *The Bell Tower* 51, no. 5 (September-October 1993): 20-21.

"How the General Grant Bell Became the Symbol of The American Bell Association" as told by Mrs. Edgar Littman, National President, 1959. *The Bell Tower Supplement* 53, no. 3 (May-June 1995): S-2.

Kimball, Carol W. "Mother Bailey's Famous Petticoat," *The Day*, New London, Connecticut, no date.

Kleven, Blanche, and Stanley Kleven. "Some New Thoughts on Collecting Figure Bells." *The Bell Tower* 51, no. 3 (May-June 1993): 20-41.

——. "The Source and Manufacture of Our Beautiful Bells." *The Bell Tower Supplement* 47, no. 3 (May-June 1989): B-T.

Lubus, Joan. "Bell Toys." *The Bell Tower* 52, no. 4 (July-August 1994): 14-40.

McCarty, Elaine. "Hawaiian Bells Talk Story, Too." *The Bell Tower* 55, no. 1 (January-February 1997): 34-40.

McMillan, Gina, and David McMillan. "Blooming Bells—Parts I and II." *The Bell Tower Supplement* 53, no. 4 (July-August 1995): S1-S22.

McMillan, Gina. "The Precious Moments Phenomenon." *The Bell Tower Supplement* 47. no. 6 (November-December 1989): S11-S15.

Meadows, Adela. *Quimper Pottery, A Guide to Origins, Styles, and Values*. Atglen, Pennsylvania: Schiffer Publishing, Ltd., 1998.

Measell, James, ed. *Fenton Glass: The 1980s Decade*. Marietta, Ohio: The Glass Press, Inc., 1996.

Nemecek, Sylvia. "Fenton Bells." *The Bell Tower Supplement* 53, no. 6 (November-December 1995): S2-S16.

Pecor, George. "Silver and Bells." *The Bell Tower* 54, no. 6 (November-December 1996): 34-51.

Ringland, Ethe. "Sleigh Bells." *The Bell Tower* Special Issue, 1990: 3-12.

Rorem, Melva. "Cathedral Bells: Bing & Grondahl's New Year Bells." *The Bell Tower* 56, no. 6 (November-December 1998): 30-31.

Schick, R.D. "S.S. Sarna: A Mini-Biography." Special Report for Heart of America Chapter of ABA, Meeting at Auburn, Kansas, Sept. 20, 1981.

Sieber, Mary, ed. *1997 Collector's Mart Magazine Price Guide to Limited Edition Collectibles.* Iola, Wisconsin: Krause Publications, 1996.

"Sleigh Bells—A Canadian Paper." *The Bell Tower* 54 no. 6, (November-December 1996): 26-27.

Springer, L. Elsinore. *The Collector's Book of Bells.* New York: Crown Publishers, Inc. 1972.

Stafford, Kathryn. "Smoke Bells." *The Bell Tower Supplement* 50, no. 3 (May-June 1992): S2-S4.

Trinidad Jr., Al. "Carnival Glass Bells." *The Bell Tower* 52 no. 2 (March-April 1994): 18-19.

——. "Sterling Silver Bells With Teething Rings." *The Bell Tower* 51, no. 5 (September-October 1993): 14-15.

——. "Steuben Glass Bells." *The Bell Tower* 54 no. 2 (March-April 1996): 14.

——. "Wedgwood Jasper Bells." *The Bell Tower* 56, no. 3 (May-June 1998): 18-20.

"Types of Glass and Their Manufacturers." *The Bell Tower* 47, no. 1 (January-February 1989): 22-26.

Index

ABA. *See* American Bell Association
alloys, 22
altar bells, 65, 66
American Bell Association, 17, 40, 49, 93, 94, 188
American Bell Association Organizational Bell, 188
animal bells, 16, 28, 60, 68-74, 124, 125
    camel bells, 72
    cattle bell, 70
    cowbells, 68-70, 87
    elephant bells, 73, 74, 124, 125
    horse bells, 70, 71
    reindeer bell, 73
    sleigh bells, 16, 68, 71
    water buffalo bell, 60
ankle bracelet, Indian, 16
Antoinette, Marie, 112
Archbishop of Rheims, 67
Arcosanti, 20
Arsklokke. See Year Bell
Art Nouveau, 25, 29
artists
    Ayres, Samuel, 47
    Barnett, Gordon, 143, 145, 146
    Buehler, Wilhelm, 154
    Burton, Frances, 40
    Chase, Lynn, 122
    Corbett, Bertha L., 52
    Daley, Betty, 126
    Davis, Lowell, 121
    De Grazia, Ted, 175
    Ele, 95
    Greiner Jr., Max, 95
    Hagara, Jan, 170
    Hanushevsky, Daria, 52
    Holgate, Jeanne, 159
    Joplin, Janice, 110
    Lamb, Robert, 188
    Lynd, Jacqueline, 133
    Mager, Kurt, 94
    Mayer, Terry, 143
    McCombie, John, 96, 102, 121
    Perillo, Gregory, 140
    Pollard, Audrey, 110
    Procopio, Adolfo, 97, 105
    Riccio, Daniel J., 94
    Ricker, Michael, 94
    Rockwell, Norman, 39, 44, 62
    Seibert, Homer B., 60

    Soleri, Paolo, 20
    Spindler, Robin, 43
    Techenor, James, 58
    Tiernan, Tim, 47
    Wiindblad, Bjorn, 34
Bailey, Anna Warner, 89
Barton, Hiram, 71
Barton, Hubbard, 71
Barton, William, 71
Bassett, Sally, 114
Bell Collector's Club, 50
bell ringers, 83
Belleek, Irish, 160
Bells of Christmas Series, 140, 141
Bells of Sarna, 123
Bells of the World's Great Porcelain Houses, 50, 161, 164
Belvoir Castle, 39
Bicentennial bells, 49, 91, 92
Bohemian bells, 8, 35, 155
Boston Tea Party, 90
call bells, 75
castor sets, 181-183
celluloid, 82, 84
ceramic bells, 7, 17, 21, 48-54, 61-63, 85-88, 90-92, 95, 106-111, 116, 117, 121, 122, 126-142, 150-153, 158-166,169-175
cherubs, 24, 49, 143, 175
Childhood Treasures Series, 171
chimes, 20
Chinese bells, 9-11, 23, 56, 57, 73, 116, 117, 121, 149
church bells, 22, 25, 89
clappers, pictured, 19, 24, 40, 49, 51, 56, 57, 59, 62, 64, 69, 72, 94, 95, 99, 102-104, 106, 108, 113, 115, 122, 124, 138, 160, 170
cloisonné, 55, 58, 144, 147
coal, 8
Collecti Bells, 42
Communion Bell, 109
Connoisseur Collection, 44, 154
cowbells, 5, 49, 68-70, 87, 146, 153. *See also* animal bells
Critter Bells, 116
crotals, 16, 17, 68, 74, 81, 82, 125, 126, 129, 131, 146, 147
definitions
    basse-taille, 55
    bisque/biscuit, 48

    brass, 22
    bronze, 22
    carnival glass, 32
    casting, 22
    champlevé, 55
    chimes, 20
    cloisonné, 55
    earthenware, 48
    figural bells, 95
    figurine bells, 95
    gongs, 9
    mechanical bells, 18
    Millefiori, 32
    open mouth, 7
    patina, 22
    plique-á-jour, 55
    porcelain, 48
    repoussé, 55
    stoneware, 48
    windbells, 20
Designer Series, 43
Dickens, Charles, 103
dinner bells, 24, 29, 62, 117, 125, 161
Don Quixote, 96, 103
doorbells, 79, 80
Earl of Mt. Batten, 15
El Camino Real bell, 179
enamel bells, 23, 55-57, 61, 144, 147, 155
English bells, 12, 15, 26, 31, 38, 39, 50, 54, 61, 89, 90, 98, 114, 135, 139, 141, 145, 150, 153, 160-162, 164-166, 178
Evangelist bells, 66, 67
Family Signature Series, 40
Fenton, Frank Leslie, 39
figurals, 23-25, 36, 41, 56, 63, 64, 67, 73, 85, 86, 88, 92, 94, 96, 97, 100-103, 105, 114, 115-122, 128-130, 136-139, 148-151, 154, 167, 168, 170, 172
figurines, 53, 62, 98-101, 103-114, 130, 134, 136-139, 167, 168, 172, 174, 181
Four Seasons bell, 49, 50
Fra. Junipero Serra, 115
Franklin, Benjamin, 91
Freedom Bell, The, 92
French bells, 19, 50, 51, 65, 67, 106, 166
General Grant Bell, 92
German bells, 32-34, 49, 52, 134, 135, 140, 149, 152, 153, 162, 165, 169
glass bells, 8, 32-47, 61, 63, 64, 87, 90, 111, 118, 129, 134, 139, 141, 154-158, 171, 173,

175, 176, 186, 187
  carnival glass, 39, 41, 46, 134
  Millefiori glass, 36
  ribbon glass, 36
glazing, 48
gnomes, 108
gongs, 9-15, 27
Grant, Ulysses S., 92
Gregory, Mary. *See* Mary Gregory
hallmarks. *See* marks
hame bells, 71. *See also* animal bells
hat button bells, Mandarin, 57
holiday bells, 60, 126-142, 176
Holly Hobbie™, 136
Hummel, 169
Incolay® Stone, 64
Indian bells, 16, 25, 73-75, 123-125, 132, 155, 167
Irish bells, 15, 38, 53, 160
Irish Dresden, 53
Italian bells, 24, 25, 36, 37, 151, 158, 159, 168
jasper, 38, 49, 50, 135
Jefferson, Thomas, 91
jewelry, 16, 28, 55, 143-147
Jones, John Paul, 91
Jones, Jenny, 114
Jones, Mary. *See* Jones, Jenny
Kentish Maid, 99
Kiwanis International, 90
lamesary bell, 66
Liberty Bell, 177, 179
Limoges, 50, 51
Lincoln Imp, 96
Lind, Jenny, 112, 113
Lucky Bells of San Michele, 147
Lucy Locket bell, 104
Madame Pompadour, 113
Mandarin hat button bells. *See* hat button bells, Mandarin
manufacturing companies and potteries
  Alfaraz Workshop, 107
  Art Foundry, 94
  Artaffects, 140
  Autumn Treasures, 131
  Avon, 136, 151
  Bing & Grondahl, 142
  Boston & Sandwich Glass Works, 173
  Boyd Art Glass Company, 111
  Buckleigh Moorland Pottery, 54
  Cambridge Glass Company, 46, 47
  Carrigaline Pottery Co. Ltd., 53
  Cavan Company, 90
  Clonmel Pottery, 54
  Coalport, 54, 141, 164
  Corning Glass Works, 47
  Danbury Mint, 8, 50, 91, 97, 138, 154, 159, 161, 164, 168
  Ebeling & Reuss, 51
  Enesco, 109, 130, 172
  Fenton Art Glass, 39-46, 89, 90, 129, 154, 156, 157, 171, 173
  Fostoria, 46
  Franklin Mint, 154, 159
  Galway, 38
  Goebel, 34, 169
  Golden Crown, 162
  Gorham, 29, 50, 62, 108, 139
  Hallmark, 126
  Hammersley, 90, 163
  Hampshire Pewter, 188

Heisey Glass Company, 46
Herend, 152, 163
Hutschenreuther, 152, 165
Imperial Glass Company, 46
Intagloi Designs Ltd., 175
Iron Art Company, 103
J.E. Caldwell Company, 89
Jasco, 116, 174
Lefton China, 131, 151
Lenox, 128
Lincoln Mint, 67, 105, 138
Lindner, 134
Lladro, 109
Lunt's Silversmiths, 126
Meissen, 162
Meriden B. Company, 18, 182, 185
Minton, 153
Niles and Strong, 71
R.S. Prussia, 51, 163, 165
Reed & Barton, 105, 137, 168
River Shore Ltd., 44, 62
Rosenthal, 34
Royal Albert, 160-162
Royal Bayreuth, 52
Royal Crown Derby, 164
Royal Doulton, 39, 161
Royal Worcester, 141, 165
Russ, 132, 166
Sandstone Creations, 175
Schmid, 90, 121
Steuben. *See* Corning Glass Works
Taylor Bell Factory, 145
Tyrone Crystal Company, 38
Viking Glass Company, 46
Villroy & Boch, 140
Waterford, 129, 141
Webster Sterling, 31
Wedgwood, 38, 49, 50, 135, 162, 171
Western Potteries, 21
Westland Company, 128
Whiting Sterling, 30
Wilcox Silverplate Co., 183
marks, 28, 30
marks, pictured, 34, 142
marriage cup. *See* wedding cup
Mary Gregory, 173
masseurs' bell, 25
mechanical bells, 18, 19, 75-80, 83, 84, 116, 178, 181, 183-185
  tap bells, 18, 19, 75-78, 178, 181, 183-185
  twist bells, 18, 75, 78-80
Merri-Bells, 174
metal bells, 7, 9-20, 22-31, 61-85, 87-90, 92-105, 112-127, 131, 132, 138, 139, 143-150, 155, 167, 168, 177-185, 188
Mexican bells, 21, 28, 54, 106, 107, 111, 140, 150
Mickey Mouse, 136
mission bells, 177, 179
monastery, 67
Mother Bailey. *See* Bailey, Anna Warner
Mother Hubbard bell, 103
Mr. Micawber, 96, 103
National Bell Collectors' Club, 92
New England Collectors' Society, 137
New Year's bells, 49
Occupied Japan, 84
Oriental bells, 12, 23
ornaments, 16, 137-139
Our Lady of Lourdes, 67

Pan, 96
patina, 22, 25, 66, 104, 112
Peanuts™, 137
Peter Rabbit, 171
Poe, Edgar Allen, 5
Popular Bell, The, 78
Potter, Beatrix, 171
powder box, 181
Precious Moments™, 172
Queen Elizabeth I, 112
Queen Elizabeth II, 38, 89
Quimper, 106
rampant lion, 122
rattles, 16, 81, 82, 83
religious bells, 14, 25, 65-67, 97
Renaissance bell, 24
repoussé, 31, 55
Revere, Paul, 91
Ross, Betsy, 91
Rothschild bird, 152
Santa Claus, 129, 134, 137-139, 176
Sarna, S.S., 72, 79, 123-125, 127, 177, 178
scrubber bell, Indian, 25
shark's mouth, 14
shells, 19, 78
slave bell, 28
slave call bell, French, 19
sleigh bells. *See* animal bells
smoke bells, 186, 187
Sonia Rose, The, 159
spaghetti ware, 53
spoon holders, 181, 184
spoons, souvenir, 177, 179
St. Joseph, 101
street car signal bell. *See* trolley bell
Sunbonnet Baby bell, 52
Swarovski crystal, 63
tea bells, 5, 22, 29, 31
teething rings, 81, 82
thimble holders, 60, 106
tiger claw, 73
Tinkerbell, 96
toe ring, Indian, 16
toys, 5, 81, 83, 84, 148
trolley bell, 27
Upsala Cathedral, 142
wani guchi. See shark's mouth
Washington, George, 91
waste bowls, 181
wedding cup, 180
windbells, 20, 21
windmill, 23, 49
wooden bells, 58-60, 61, 62, 70, 72
World's Fair, 92
Year Bell, 142
Ziggy™, 137

About the Author

Donna Baker is a writer and editor from West Chester, Pennsylvania. A member of the American Bell Association International, Inc., she is also the author of *Collectible Bells: Treasures of Sight and Sound.*

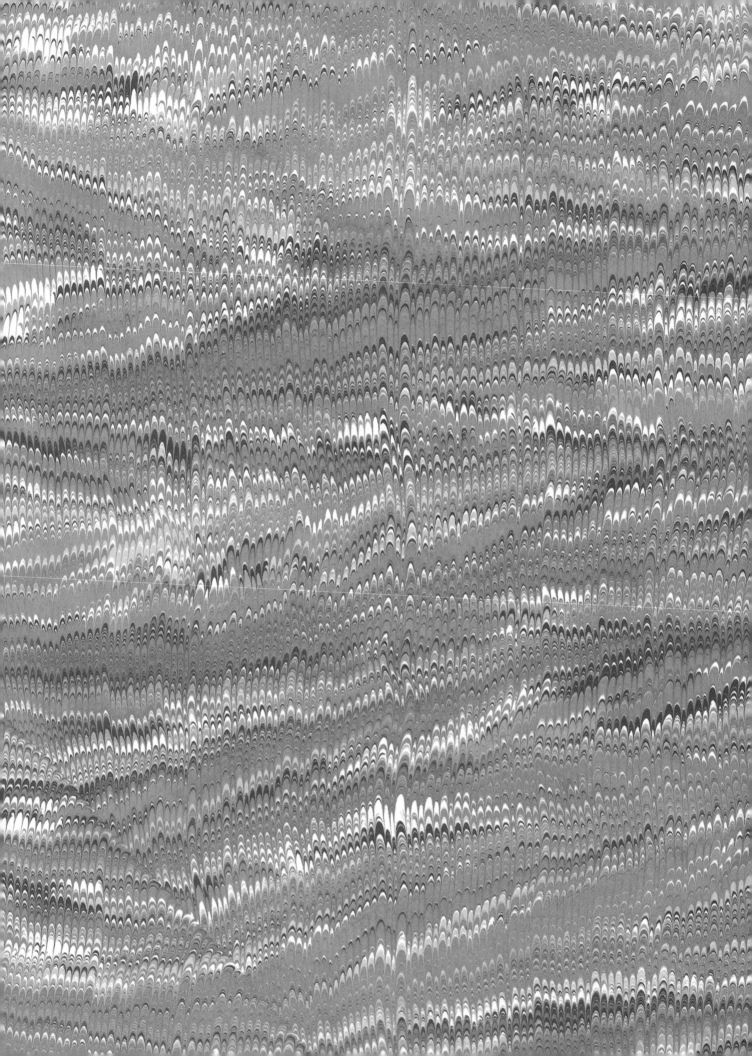